T0298428

Amphibians and Reptiles of North-west Europe

Amphibians and Reptiles of North-west Europe

Their Natural History, Ecology
and
Conservation

Ian F Spellerberg

Illustrations by
Peter Jack

Science Publishers, Inc.

Enfield (NH), USA   Plymouth, UK

SCIENCE PUBLISHERS, Inc.
Post Office Box 699
Enfield, New Hampshire 03748
United States of America

Internet site: *http://www.scipub.net*

sales@scipub.net (marketing department)
editor@scipub.net (editorial department)
info@scipub.net (for all other enquiries)

ISBN-13: 978-1-57808-259-9 (pbk)

Published by Science Publishers, Inc., Enfield, NH, USA

Foreword

There is a proverb to the effect that 'the frog cannot get out of its pond'. This is usually taken to imply the futility of altering or concealing one's true character. It could however equally well reflect the inability of people to abandon their deeply held interests. So, it seems appropriate that I should mention this in a Foreword to Ian Spellerberg's absorbing and finely illustrated book. Under professional guidance of such quality it is all too easy to become attracted by the biology and ecology of amphibians and reptiles. Once this fascination has been acquired, it may be difficult to abandon it. Let the casual reader therefore be warned: you could well find that herpetology grabs you like Virgil's 'snake that lurks in the grass'. Before you know where you are, you may discover that you are a dedicated herpetologist.

Anything which is worthwhile does require effort and hard work before it can be enjoyed and appreciated to the full. If it is worth doing it is worth doing well! Herpetology is no exception. So if, like the frog in its pond, you would yourself 'a wooing go' and pursue this beguiling new interest, you should do the job thoroughly. Commence your studies with the help of this book that is not merely a field guide, but which gives up-to-date accounts of the ecology of each species. Such knowledge is so important to the conservation of these species. If the amphibians and reptiles of the temperate parts of Europe are not well managed for their conservation, there will be serious losses. Such losses will be a poor reflection of the 'health' of our environment.

Although Ian Spellerberg has spent many years researching the ecology and conservation of amphibians and reptiles, he has acquired world-wide experience of many other aspects of ecology and conservation. In this his latest book, he has adopted a somewhat unusual approach and gives not only details of taxonomy, distribution, reproduction, growth and development, but also valuable information such as environmental adaptations and numerous aspects of the ecology of the various species discussed. The fine illustrations by Peter Jack are better than any photograph.

This book will be extremely useful to conservationists; amateur herpetologists and others concerned about the decline of our amphibians and reptiles. The text is free of obscure jargon and is written in a way that any lay reader, without specialist knowledge of biology, will be able to understand and enjoy. At the same time, this book is sufficiently detailed to be a valuable reference work for undergraduates and research students.

J.L. Cloudsley-Thompson. MA, PhD, DSc, Hon DSc, Hon FLS, FIBiol, FWAAS,
Emeritus Professor of Zoology, University of London.
Past President and Hon. Member of the British Herpetological Society.

Preface

This book provides brief accounts of the natural history, ecology and conservation of a few amphibian and reptile species from western Europe.

Amphibians and reptiles are much maligned animal compared to other groups. This is perhaps one of the reasons why they have been relatively less well conserved and protected. Many are highly endangered and there are concerns about the overall decline in amphibians throughout the World. It has been suggested that this decline is an indication of a deteriorating environment.

Conservation of wildlife depends on many things. Education, legislation and protected areas are just a few of the dimensions of conservation biology. In addition, natural history and ecological studies have long contributed to conservation and management of wildlife.

It is perhaps fortunate for conservation of European wildlife that Europe has been rich in natural history studies for a very long time. Ecological science emerged from that rich tapestry of natural history and there is now a wonderful legacy of information on some forms of European wildlife.

The level of knowledge about European amphibians and reptiles is nevertheless very patchy with some species being the focus of much attention while we know very little about other species. This has become very clear when the text of this book was prepared. While focusing on the natural history, ecology and conservation of amphibians and reptiles in temperate Europe, it became clear that some species have not been well studied in the field.

The illustrations in this book are a study in natural history in their own right. Indeed paintings and drawings are an integral extension of natural history. Ecological field studies are more than just going out and measuring things. It takes a dedicated and skilled person to be very observant and be able to record and illustrate what has been observed. Perhaps educational programmes about field studies could pay more respect to our wonderful legacy of natural history diaries and illustrated records.

It is hoped that this book will prompt many more dedicated studies of amphibians and reptiles as well as providing a basis for conservation.

Ian Spellerberg,
Director, Isaac Centre for Nature Conservation,
Lincoln University, New Zealand
Visiting Professor at the University of Southampton, U.K.

Acknowledgements

This book could not have been published without generous grants from the Linnean Society for the preparation of the illustrations and sponsorship from British Petroleum Exploration.

The content of the book includes a personal selection of species and has a personal bias in terms of the topics. The responsibility for any errors and mistakes is my own. My gratitude is extended to the many people who have helped me in this project. Peter Jack showed much tenacity and perseverance during the preparation of the illustrations. Nick Arnold (British Museum, Natural History) very kindly provided me with much information and advised on the illustrations. Nick Smith (University of Southampton) and Larry Mortlock (Lincoln University) kindly helped to prepare the distribution maps. Miss J. Welch (previously University of Southampton) generously helped with translations. Jane Swift (Lincoln University) provided much needed editorial skills for the manuscript. Michael Packard helped to improve the text.

The following have kindly contributed in some way to this project: Claes Andrén, Nick Arnold, Roger Avery, Robert Barbault, J. Barandun, Trevor Beebee, Thomas Böker, Florentino Braña, Peter Brown, Pedro Galán, Tony Gent, Dieter Glandt, Paul Jutkiewicz, Göran Nilson, Chris Reading, Ramon Saez-Royuela, Anton Stumpel and Michael Warburg.

The Polish Authorities kindly gave their permission to illustrate a set of postage stamps from Poland and the American Association for the Advancement of Science gave their permission to use material from the Journal, *Science*.

Contents

Introduction

Herpetology, the study of amphibians and reptiles, is attracting more and more interest throughout Europe but, unfortunately, the animals themselves are generally perceived as unattractive. In contrast, furred, and especially feathered, vertebrate animals are more popular so that they encourage their own specialised societies such as the Royal Society for the Protection of Birds (RSPB). The unhelpful attitude of many people towards amphibians and reptiles can be attributed to many fallacies down the ages, but more education and information about them is slowly changing the picture. The world's amphibians and reptiles are part of a diverse and remarkable group of animals, and the comparatively few species found in Europe have been considered sufficiently beautiful to depict on postage stamps (Fig 1). Much has certainly been written about them.

Books about amphibians and reptiles of Europe

One of the largest volumes was written by Roessel von Rosenhof and published in a large format (45 x 35 cm) in 1758. Another early book by Schreiber, published in 1875, provided much detailed morphological information. Other early volumes concentrated on particular taxonomic groups. A good example was Latreille's on Salamanders with very fine coloured illustrations, published in 1800. Yet another with beautiful pictures was by Cooke on British reptiles, which appeared first in 1865 and a later edition in 1893. Full references are given at the end of the book. In more recent times there have been many kinds of books on European herpetology, some again concentrating on a single taxonomic group such as Steward's very readable *The Snakes of Europe* (1971). Others like Arnold and Burton's *Field Guide to the Reptiles and Amphibians of Britain and Europe* (1978, reprinted 1993), are excellent sources of information and useful field guides. More recently there is Richard Griffiths book *Newts and Slamanders of Europe* (1996).

The variety is impressive, ranging from the small, very good pictorial introduction, Ellis' British Amphibians and Reptiles (1979), to the very comprehensive books on single species, Beebee's *The Natterjack Toad* (1983) or Nollert's *The Spadefoot Toad* (Die Knoblauchrote) of 1984. Other notable and definitive volumes are Merten and Wermuth's *Die Amphibien und Reptilien Europas* (1960), Andreas and Christel Nollert's *Die Amphibien Europas* (1992), Gruber's *Die Schlangen Europas* (1992) and in particular the recent, and much welcomed, volumes of the *Handbuch der Reptilien und Amphibien Europas* edited by Wolfgang Bohme. The number of herpetological books in German is indicative of a long tradition of natural history and herpetology in Germany. Books on the amphibians and reptiles of Britain alone include Malcolm Smith's classic *The British Amphibians and Reptiles* published in several editions, Simm's *The Lives of British Lizards* and the more recent book by Beebee and Griffiths.

Research reports and publications

The quality and ecological importance of research on amphibians and reptiles of Europe has increased steadily over the last few years. Nevertheless, it is becoming increasingly difficult to obtain support for this research, and it appears that a disproportionately small amount of dwindling funds is directed towards projects on these animals.

Results of amphibian and reptile research may appear in the form of student theses, papers in scientific journals or contributions to symposium volumes. Despite the odds against them, the number of M.Phil. and Ph.D. theses on amphibian and reptile ecology have greatly increased in recent years but, unfortunately, too little of that material – with notable exceptions – has appeared in the scientific literature.

Research published in symposium volumes has been important and Rocek (1986) and Glandt and Bischoff (1988) edited two excellent examples. Comparative ecological studies in scientific journals have particularly important implications for conservation and management, and examples include Strijbosch (1979, 1980), Beebee (1985), Arnold (1987) and Massot et al. (1992) – see the references at the end of this book.

Other examples of scientific papers giving results from research on amphibians of Europe have been important for conservation and management. These include van Gelder et al. (1978) on calling behaviour of frogs, Griffiths (1987) on niché studies of newts, Halliday and Arano (1991) on the phylogeny of newts, Harrison et al. (1984) on morphometric observations, Kalezic and Hedgecock (1980) on genetic variation in newts, Lizana et al. (1990) on community structure of amphibians, Mann et al. (1991) on the importance of habitat complexity for amphibians, and Reading's research on the breeding biology of toads (see for example, Reading and Clarke, 1983). The reptiles described in this book have been the subject of some very detailed and useful ecological research relevant to conservation. Some of the research was undertaken at the University of Southampton and includes the work of Nicholson (1980), Goddard (1981), House (see the account of the Sand Lizard), Dent (1986), Gent (1988), Gaywood (1990), Smith (1990) and Brown (1991).

About this book

The questions therefore arise: why write another book about these animals, and why make this selection of amphibians and reptiles? The urgency to undertake good conservation measures for many of Europe's amphibians and reptiles is without question but it appears that conservation efforts have not made the best use of a wealth of published research. Therefore the main reason for writing this book is to promote the need for conservation based on good ecological research, and especially on good fieldwork. In the past, many books on amphibians and reptiles have included accounts of the biology of various species but often that information is out of date or lacks ecological perspectives. The synoptic accounts used in this volume makes as much use of ecological information as possible. They are drawn from a wide range of scientific papers where studies have been field based or undertaken in large outdoor vivaria. In addition, we have drawn on information kindly provided by my colleagues. Myself and Peter Jack have endeavoured to see all species described here in the wild and to undertake studies on most. The length of the species accounts therefore reflects the amount of research results available for study.

There are three concerns expressed throughout the book: the practice of keeping wild animals in captivity; the importance of field studies; and under-used research results for management and conservation. It is not easy to come to terms with taking amphibians and reptiles – or birds or mammal – from the wild and keeping them in captivity; and certainly collecting them simply as a hobby is very difficult to justify. Fortunately there is a move away from this practice. By way of contrast there is much to be done to encourage good field studies, especially protracted ones. Little is known about long-term changes in amphibian and reptile populations, and it is only by carefully recorded studies that we can begin to understand why there are fluctuations in numbers. Ecological research over the last 50 years has produced much information on the amphibians and reptiles of Europe but still only a small proportion is used for management and conservation. Conservation organisations should take note.

The selection of species for the text

As with other animals, the ecology and distribution of amphibians and reptiles has been determined mainly by climate and vegetation. There is an increase in species richness in European amphibians and reptiles from north to south which can be explained, in part, by the changes in climate and vegetation patterns that have taken place over the last 10,000 to 12,000 years, after the last glaciation. The present patterns of vegetation and climate in Western Europe form bands, which range from the broadleaf and coniferous forests of the north to the subtropical areas of the Mediterranean. It is possible to identify a number of major climatic regions such as the Arctic, Subarctic, Atlantic, Continental and Mediterranean (there is more than one classification of climatic regions and each is based on certain climatic and other variables).

My own research has been largely on ecology and conservation of amphibians and reptiles of temperate climatic regions in various parts of the world, and I have always been interested in how these animals have adapted to cool and cold climates (see for example Spellerberg 1976). The basis for the selection of species in this book has therefore been biogeographic and not by country. I have chosen species whose distribution occurs, at least in part, in the Atlantic climatic region. (Fig.2) That region is part of the temperate forest biome where temperature extremes are lacking and warm summers have at least four months with temperatures greater than 10° C. Interestingly enough, the main areas of lowland heathlands of Europe are found in the Atlantic climatic region and, of course, the heathland or ericoid type of plant community is important for many amphibians and reptiles.

The species accounts

These descriptions are meant to be brief synopses of ecological information. They are not definitive and for reasons of space I have had to be selective. When describing the species, certain conventions have been used. For example, the terms dorsal (upper side or back) and ventral (underside of belly) have been used. Lengths of species are given and may often be the length of the animal from the tip of its snout to the vent or cloaca (snout-vent or s-v length); it is used for reptiles because many individuals lack a full undamaged tail. Other terms have been defined in the Glossary.

The scientific name is followed by the vernacular or common names in English (GB), French (FR), German (DE), Dutch (NE), Spanish (ES) and Swedish (SE).

This book can be used as a field guide but is intended to generate interest in the ecology and conservation of these species and to introduce the reader to some of the more interesting aspects of ecological research. I have therefore included important references at the end of each species account as well as a general reading list at the end of the book. But let us not forget that modern ecological studies owe much to good natural history in the past. As a tribute to previous natural historians I have included a few short delightful quotes from the old literature.

Paintings of the species rather than photographs have been used because certain taxonomic features can be emphasized by this medium to aid identification. There are distribution maps for all species (except tortoises, terrapins and turtles) but their accuracy is limited as they can give only a very general indication of the broad geographical distribution of a species.

The next section of the book is a general introduction to the biology and ecology of the amphibians and reptiles, and not just about those found in Europe. There then follow the species accounts and finally some case studies in conservation are provided.

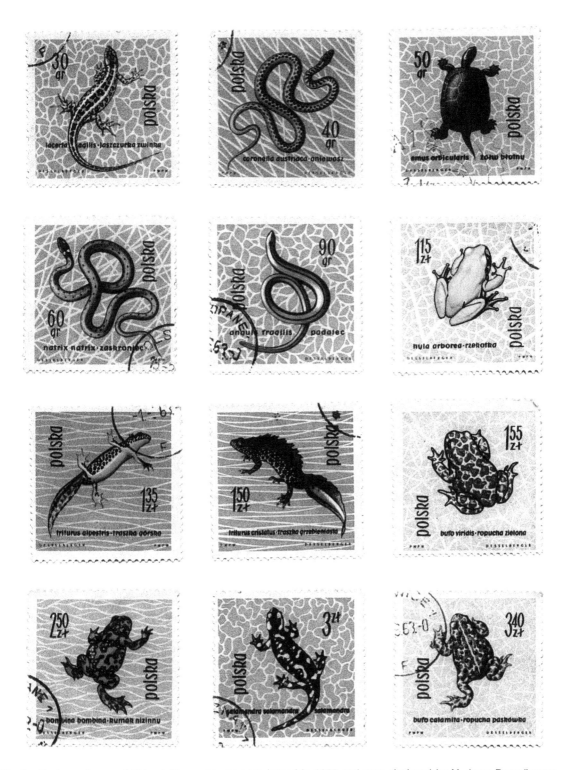

Fig 1. Postage stamps of Poland. These stamps were issued in 1963 and were designed by M. Jerzy Desselberger. Reproduced with kind permission of the Director of Postal Services, Poland.

Fig 2. The Atlantic Climatic Region of Europe (an oceanic type of climate, lacking temperature extremes and abundant, well distributed rainfall). To the East is the Continental Climatic Region, to the North the Subarctic Climatic Region and to the South the Mediterranean Region. From Spellerberg (1976).

The Ecology of
Amphibians and Reptiles

Preamble

This chapter considers the world's amphibians and reptiles and not just those found in Europe. Their extinct and living forms have been the subject of much fascination to many generations but the study of them (herpetology), up to about the mid-twentieth century, concentrated more on morphology and anatomy. Many early natural history books contain captivating accounts of amphibian and reptilian behaviour and, although often inaccurate, they are worth mentioning occasionally because they were written with so much charm and enthusiasm. It has only been over the last 50 years of the century that there has been a marked increase in the number of serious field studies and ecological research. This research has been dependent on advances in survey methods (see for example Foster and Gent, 1996). More ecological and behavioural studies, especially field studies, on amphibians and reptiles have greatly advanced our knowledge of these animals, and they have helped to dispel many fallacies and to be more rigorous about the use of biological terms. We now know, for example, that amphibians and reptiles are not really cold-blooded, that they do not hibernate (in a technical sense of the word), and that species coexisting in the same area or pond can have highly specialized methods for avoiding competition. Most importantly, since many of these animals are now endangered species and in need of active conservation, ecological studies have a major role to play in establishing conservation measures for them.

Taxonomy and systematics

Before we can study any animal we must know what we are studying. Organisms must therefore have a name and that name should be acceptable to the world's scientific community. The use of vernacular, or common, names such as Midwife toad or Wall Lizard have a basic disadvantage because different languages or regions will have different vernacular names for the same kind of organism, and occasionally the same language will have several vernacular names for the same species, for example, Warty Newt or Great

Crested Newt in English. The use of scientific names of species follows an agreed world standard procedure. The most commonly encountered scientific name is a binomial one comprising Genus and species. In, say *Triturus alpestris*; the Genus or Generic name is *Triturus*, and the species or specific name *alpestris*. There is only one Genus *Triturus* but it may have several species and subspecies.

A species, a basic taxonomic unit, can be defined as a group of interbreeding individuals, which normally does not breed with other groups. A subspecies is a group of individuals within a species, which have characteristics that distinguish it from other groups of individuals of the same species; but subspecies can interbreed. Nothing is ever clear-cut in nature and this certainly applies to the definition of a species. In most cases we can recognize what a species is, but some species do interbreed, and that has made it difficult to determine whether there is a true or hybrid species. The frogs *Rana lessonae* and *R. esculenta* are two such forms which have attracted much discussion concerning their true or hybrid form. In nature there is so much variation that the appearance of some species can change over a geographical area. The lizard *Podarcis muralis*, for example, has many differences in its colour. The term "cline" means the gradual change in characteristics of populations (of a species) over a geographical area.

Taxonomy deals with the naming of organisms and systematics with their classification. When we classify *Triturus alpestris*, for example, we are placing it in a group containing related amphibians. Systematics has given us a very helpful, logical, hierarchical classification starting with a broad group and descending through the Genera and species.

Over the years the agreed hierarchy has been refined and various groups of amphibians and reptiles have been regrouped and separated as our knowledge has advanced. The group name 'Batrachians', for example, was once used to refer to the amphibia. Early naturalists regarded salamanders as batrachians but some went on to say "the salamander belongs to that order of reptiles – the Batrachians – which are those with naked bodies, without the hard covering of the tortoises, or scales like serpents". We can see that the groups have changed and we now put salamanders firmly among the amphibians which, in turn, are separated from the reptiles.

Today there is a fairly well accepted hierarchy of taxonomic groups. *Triturus alpestris*, for example, is in a Family of newts called Salamandridae. That Family is one of a number in the next group known as an Order. The Order is the Caudata. The next broad group is the Class Amphibia, and above that is the very broad group or Phylum, Vertebrata. In other words, 'from the top' the hierarchical classification of *Triturus alpestris* is as follows:

Phylum – Vertebrata
Class – Amphibia
Order – Caudata
Family – Salamandridae
Genus – *Triturus*
Species – *alpestris*

It is conventional to write the Genus and species names in Italics and to give the first letter of the Genus a capital letter. The species is written in lower case. That makes it easy for anyone in the world to know which is which.

There are about 4,000 species of amphibians and about 6,500 species of reptiles living today. The Class Amphibia includes the following groups:

Order Apoda (Gymnophiona), caecilians, 163 species

Order Caudata (Urodela), salamanders and newts, 358 species

Order Anura (Salienta), toads and frogs, 3495 species

The names in parentheses are alternative names for the Orders.

Caecilians are slim worm-like amphibians with no limbs and practically no tail. Most are about 25-30 cm long. They are forest animals found in Central America, Africa, India, Indonesia and elsewhere in the Tropics.

The Order Caudata, or Urodela, includes those amphibians which retain a tail throughout their lives, unlike the frogs and toads. The salamanders are the terrestrial forms and the newts the aquatic or semi-aquatic species.

There is no biological distinction between toads and frogs – the Anura or tail-less amphibians – but generally speaking the typical frogs are in the Genus *Rana* and the toads in Genus *Bufo*. One of the common features is that they tend to aggregate to breed.

The world's living reptiles are divided into the following groups:

Order Chelonia – tortoises, terrapins and turtles – 244 species

Order Rhynchocephalia – tuatara – 1 specie

Order Squamata:

    Suborder Amphisbaenia – worm lizards – 140 species

    Suborder Ophidia – snakes – 2390 species

    Suborder Lacertilia – lizards – 3750 species

Order Crocodylia – alligators and crocodiles – 22 species

There is no mistaking any member of the Order Chelonia because all have the short, wide body in a protective shell with the upper part being the carapace and the lower part the plastron. They do not have teeth but have sharp horny beaks (I use the terms as follows: birds have bills whereas Chelonia have beaks). The terrestrial forms are the tortoises, terrapins are semi-aquatic, and turtles are the marine species.

The tuatara (Figure 3) is a lizard-like reptile, about 62 cm in length, found only on offshore islands of New Zealand. It is the sole survivor of an ancient group of reptiles, which has remained unchanged since the time of dinosaurs. Although lizard-like in appearance, tuataras have several anatomical features that distinguish them from lizards.

The Amphisbaena (worm lizards) is in an equally strange group of limbless, burrowing reptiles (Figure 4) found in South America, South Africa and Spain.

Description

Amphibians have smooth, moist skins without scales. Many are very colourful, or at least have distinctive markings, and it is these markings that can sometimes be used to identify individual animals in field studies. Most amphibians lay eggs in water or in moist surroundings, and each egg is covered in gelatinous material and not a shell. Usually the eggs hatch into

Fig 3. The tuatara (*Sphenodon*) of New Zealand. A lizard-like reptile but recognized as the single species in the Reptile Order Rhyncocephalia. Adults grow to about 62 cm in length. From Spellerberg (1982).

Fig 4. An amphisbaenid, *Blanus cinereus* from Spain. This species of reptile reaches about 25 cm in length. From Spellerberg (1982).

larvae, which are morphologically different from the adult.

Sophisticated means of vocal communication have evolved among frogs and toads. The calls are used to identify individuals of the same species, potential mates and potential rivals.

Throughout the world, frogs, toads and salamanders have poisonous skin glands as a means of trying to avoid being eaten. The strength of the poison varies but, in South America, there are some frogs whose very toxic venom indigenous people use on arrow tips when hunting. Sometimes the poison glands are concentrated just behind the eyes, as are the paratoid glands of the common toad *Bufo bufo.* The great crested newt *Triturus cristatus* has poison glands on its back and tail.

Reptiles have scales and a dry skin, which is not impervious; they tend to lose body water through their skin more than through their mouths. They do not have a larval stage of development and, most importantly, their eggs have protective membranes and a shell.

Throughout the world there are two species of lizards and many species of snakes, which are venomous. In Europe only about a third of the snake species are venomous and in northwest Europe only the adder *Vipera berus* is venomous.

Snakes and lizards are by far the most common representatives of all living reptiles, and they are part of a very diverse and well-adapted group. Snakes have very specialized skulls and the two halves of their lower jaws are not fused. They have elongate bodies and no limbs, although a few species do have the remnants of pelvic limp girdles and the vestiges of hind limbs. There has been much recent debate about the significance of early fossil snakes with posterior legs (Coates and Ruta, 2000). Most lizards have two pairs of legs but a few species have lost one or both pairs, although all retain at least traces of the pectoral girdle.

Many lizard species have the ability to shed their tails (tail autotomy) to provide a means of escape from predators. Tail autotomy has two notable features: the fracture plane where the tail breaks is within a vertebra and not between vertebrae; the ability to shed the tail is more likely to occur at high rather than low temperatures. The lizard can re-grow another tail, which never looks completely like the original.

Distribution and habitat

Amphibians and reptiles are found throughout the world apart from Antarctica. They are found in seas, rivers and on land as far north as the Arctic Circle and up to altitudes of 4,000 and 5,000 m in South America.

The present pattern of these animals' distribution in Europe has been largely determined by land use, particularly during the last few hundred years. Previous distribution patterns were mostly governed by post-glacial events going back many thousands of years. Following the northern retreat of the last ice age, amphibians and reptiles were able to move northwards too, culminating in an interesting distribution when the English Channel formed about 7,500 years ago (Figure 5).

Another factor leading to extended distribution has occurred when some species have

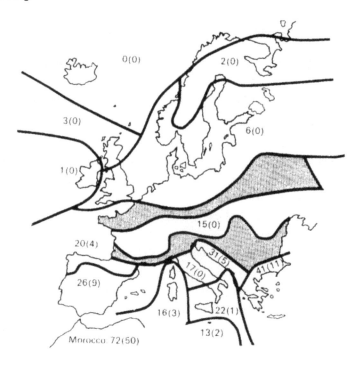

Fig 5. The numbers of reptile species (excluding tortoises and introduced species) in Europe. The first figure is the total number of species and the figure in brackets is the number not found elsewhere. The shaded areas are regions where the pattern is less clear. From Spellerberg (1982).

been accidentally or deliberately released outside their normal geographical areas. In Britain, for example, Tree frogs (*Hyla*) were introduced to the south of England. On the Isle of Wight and elsewhere populations of Wall Lizards *Podarcis*, and even the Africal Clawed Toad *Xenopus*, now exist. Introductions are not always popular: in Germany, for example, the American Bullfrog, together with other exotic amphibians, has been declared an unwelcome alien.

Recording, mapping and monitoring the distribution of amphibians and reptiles has become a very important part of their management and conservation. Recording and mapping is undertaken at many different scales from country to continent and results published in *The Atlas of European Amphibians and Reptiles* (published by the Societas Europaea Herpetologica and produced by the Service de Patrimonie of the Museum National d'Historie Naturelle, Paris) are a tribute to the dedication and enthusiasm of those interested in herpetology. The methods for monitoring amphibians in Britain have been assessed by Oldham and Nicholson (1986) and Swan and Oldham (1989). They found that nighttime counts with the aid of torches are highly variable. Although trapping with immersed traps suffers from the fact that there is a peak catch about one month after the first one, it is a more reliable method than torch counting. Above all it is important that repeated observations are undertaken using standardized methods, and the ancillary information on weather and pond conditions are also systematically recorded.

Seasonal behaviour

The Atlantic Climatic Zone has regular, contrasting seasons and therefore the amphibians and reptiles have seasonal activities. Their timing may vary within the geographical range of species. So in the species accounts that follow generalizations have to be made (but interesting exceptions are noted).

Many undergo winter topor (not hibernation in the technical sense of the word – see Glossary), followed by emergence during the spring and breeding. Autumn is the important time for many species when fat supplies are built up prior to the entry to winter topor. Seasonal movements can be quite striking when some species migrate from wintering to breeding grounds. The tenacious use of migratory routes has caused some conflicts where roads and other barriers have been built across them. There have now been some interesting developments, such as underpasses, to assist migratory movements.

Vagility and population ecology

Vagility is the term given to movement in a spatial sense. Studies of vagility include home ranges, territories and distances moved from day to day. A territory is an area, which is defended by an individual, and few species of amphibians and reptiles have them. A home range is the area in which the animal undergoes its normal behaviour; it can sometimes be calculated using the distances moved by an individual over a period of time. The number of measurements affects the accuracy of the calculation, and a correction factor may be used when only a small number of measurements have been obtained. In the following species accounts the home range area is given as a corrected or uncorrected value.

Population ecology is the study of the structure of populations: what the age classes are, for example, when they breed, or how long they live. There is much variation in life span amongst amphibians and reptiles. Some frog species will live on average for only two or three years, whereas some snakes may have a longevity of 15 or more years. Long-lived species tend to have a low level of reproduction whereas short-lived species tend to have a high reproductive level.

Feeding ecology

Studies of feeding ecology have helped our attempts to understand how two or more species can coexist without being in direct competition. In nature if two species do compete for resources such as food, one species will eventually disappear. Studies of how species use food and space have led to the concept of a niché. A popular definition of a niché is 'the activity range of each species along every dimension of the environment'. The dimensions of the niché include food types, space and time of activity. For example, two species of newts living in the same pond may avoid total competition when one species feeds on slightly larger prey or early in the morning, while the other feeds on small prey or during the late evening.

In some parts of the world, such as Central America and Australia, many species of lizards share the same habitat and resources in very subtle ways. Several species of Anolis lizards in Central America, for example, can coexist because they feed on different prey and use different parts of trees and shrubs. In temperate Europe some of the most exciting studies on feeding ecology have been carried out on newts.

Thermal ecology

The term 'cold-blooded' is old-fashioned and misleading and tells us nothing of the thermal ecology of amphibians and reptiles (Cloudsley-Thompson, 1999). 'Poikilothermic', meaning of variable temperature, is also unhelpful. By way of contrast the terms ectothermic and endothermic are useful and have come from 40 years of research on thermal ecology and energy metabolism. Most species of amphibian and reptile regulate their body temperatures largely by using heat from the environment (ectothermic), whereas birds and mammals maintain their body temperature levels as a result of internal heat produced by metabolism (endothermic). Some species of amphibians and many species of reptile are able to regulate their body temperatures in a precise manner, largely by behavioural means. Most importantly, many forms of behaviour are temperature dependent: for example, visual acuity, efficient digestion, vocalizations and speed of movement all depend on certain body temperature levels.

Hibernation is a specialized, physiological state found in some small mammals during winter. Amphibians and reptiles do not make these specialized physiological adaptations to winter but they do go into a state of topor. During the over-wintering period or period of time in topor, very little energy is used.

Regulation of body temperature, which occurs amongst some amphibians and many reptiles, is achieved mainly through their behaviour. Some species have a spectacular array of basking postures while others cleverly alternate between shady and sunny areas. There is a popular misconception that darker-coloured amphibians and reptiles will absorb heat more readily than lighter-coloured forms. The visible colour of the skin has very little to do with the rate of heat uptake. The texture of the skin and its ability to absorb or reflect infrared light (which we cannot see) is the important factor. For example, we might expect two frogs, both green in colour, to absorb or reflect heat in an equal manner. It may not be true as shown in Figure 6.

Studies on the thermal ecology of amphibians and reptiles has resulted in a terminology which, although common in the scientific literature, is not yet widely used in general or popular texts. For example, the amphibian or reptile body temperatures at which normal activity occurs are called voluntary or preferred temperatures. If an amphibian or reptile is cooled (in artificial surroundings), its body temperature will eventually reach a low level at which the animal can no longer move about. That low body temperature is called the Critical Minimum temperature. In the wild, amphibians and reptiles will try to avoid conditions

Fig 6. Two species of North-American frogs photographed with ordinary film (above) and infra-read film (below). Although visibly green, the two species have different reflectance properties (visible colour is not a good guide to thermal characteristics of amphibians and reptiles). Reproduced with kind permission of the American Association for the Advancement of Science. Source: Schwalm, P.A. et al., 1977, Science 196, 1225-1227.

that could lead to their body temperatures reaching a level where mobility ceases. Both the voluntary and the critical minimum temperature levels decrease slightly in winter and increase slightly in summer – seasonal acclimatization processes.

There is also an interesting twist in the tail when it comes to the thermal ecology of reptiles. The temperature of incubation can affect the outcome of the sex of the reptiles, particularly amongst Crocodilia and Chelonians. It has been found, for example, that some turtle species have more females after high temperature incubation of the eggs and more males after low temperature incubation. This pattern may be the reverse for other species.

Reproduction

All amphibians lay eggs which lack outer protective shells. Different amphibian species have remarkably different breeding strategies, and the variation in reproductive modes is greater than in any other group of vertebrates. There are some amphibians, such as the Common Toad *Bufo bufo*, which are 'explosive breeders' when many individuals in a population breed together over a very short space of time. Other species, such as *Bombina variegata*, have a prolonged breeding period with several short spawning periods. We need to look to the environment in which they live to find the explanation for these differences. In some cases the environmental conditions are fairly stable and/or predictable, whereas in other localities the opposite is so when the best breeding strategy might be to extend the breeding season over several months.

Comparatively little attention has been paid to the larval stages of some newts and tadpoles, and consequently their ecology and behaviour are not well understood. In general, amphibian larvae have gills and gradually change (metamorphose) into adults which have lungs. Sometimes, though, the larvae do not metamorphose but continue to grow into large, larval forms, caused sometimes by environmental conditions. This phenomenon is called neoteny. In some species adults continue to look like larvae complete with external gills (paedomorphosis). There is also a very interesting variation in the reproductive ecology and behaviour of amphibians, with some so highly specialized that they are now faced with extinction. Many species have adapted to arid conditions so we should not think of all amphibians living in or near water.

Most reptiles lay eggs (oviparous) while some give birth to live young (viviparous). A few species of lizards actually have a basic placenta. But it is the mode of reproduction and the nature of the reptile egg, with an outer protective shell, which has allowed reptiles to diversify and occupy a greater range of habitats than amphibians.

Internal fertilization is practiced by all reptiles, when the sperm is transferred directly into the female's cloaca, a common opening that transmits both excretory and reproductive products. Fertilization in the Crocodilia and Chelonia is achieved by a single male organ, but in lizards and snakes the male organ is paired (known as hemipenes) though only one is used at a time.

There are some instances of parthenogenesis amongst lizards when some populations consist entirely of females and all members are genetically identical.

SPECIES ACCOUNTS

Family Salamandridae

Salamandra salamandra

European, Fire or Spotted Salamander (GB), Salamandre commune (FR), Feuersalamander (DE), Vuursalamander (NE), Salamandra Común (ES), Eldsalamander (SE).

Introduction

The Family Salamandridae (newts and salamanders) is the most widespread of all families of 'tailed amphibians'. They are relatively small amphibians but few could be more spectacular, at least among European newts and salamanders, than the striking black and yellow colours of *Salamandra salamandra*. There are claims that the colourful appearance of this animal has a warning function, which is possible, as it is able to protect itself with noxious skin secretions produced by large numbers of ducts from cutaneous glands. However, skin secretions are not uncommon among other amphibians, of which some species are not at all spectacular in appearance and colour.

The name Fire Salamander could have many origins. In some early books there was the explanation that, when thrown in the fire, the flames could be extinguished as a result of the animal bursting and ejecting fluids.

Taxonomy

The variation in colour and body proportions is so great that some 15 sub-species have been recognized. One example, the sub-species *S. s. infraimmaculata*, is found in Israel and is considerably larger than the nominate form *S. s. salamandra* in Europe.

Protection

This species is protected in Belgium, France, Germany, Hungary, Luxembourg, The Netherlands, Poland, Spain and Switzerland.

Description

This salamander cannot be confused with any other species in Europe and is easily recognized by its robust build and striking, colourful appearance. It is the only European species in which most adults and young have intense yellow, orange or reddish spots or stripes on a uniformly shiny black background. The adults may reach lengths of just over 20 cm (including the tail) but most are smaller. Males are usually smaller than the females. There is, however, much variation in their colour and markings: some are almost entirely yellow or orange with black stripes.

Distribution and habitat

The geographical distribution is extensive: across central and southern Europe (not Britain) from Denmark to northwest Africa and Israel, and to parts of southwest Asia. The larvae inhabit permanent streams or temporary pools and, with the adults, are almost always found in hilly and mountainous regions (up to about 2,000 m, and in one locality in the Guadarrama Sierra as high as 2,280 m). Both shade and moisture affect its choice of habitats; particularly suitable are damp wooded areas, where deciduous woodlands, especially beech woods, are favoured, possibly due to the availability of leaf litter. Some populations of the sub-species *S. s. almanzoris* have adapted to live permanently on small glacial lakes. Other sub-species have adapted to life in semi-arid habitats.

Seasonal movements and behaviour

In Central Europe this salamander is not normally active during the winter, and then only limited activity occurs from about March to the end of summer. In Spain it is not active during the summer, and its activity period is linked to humid seasons. These animals tend to be rather solitary and nocturnal but some have occasionally been found over-wintering in small groups.

Vagility and population ecology

In Europe, adults have usually been found to move only short distances but, on occasions, some individuals have been found to travel distances of 300 m. They have very limited home range movements up to about 12 m. In some localities, single animals have been found at an average of one every 63 to 84 square metres.

Feeding ecology

This amphibian is mainly nocturnal but may be active during the day following periods of heavy rain. It preys on a range of invertebrates including molluscs, crustacea, some species of worms and beetles, and also myriopods (centipedes and millipedes).

Thermal ecology

Some populations inhabit semi-arid habitats where there are long hot summers. Adults of those populations have higher thermal tolerances and lower water loss than populations from cool, moist habitats. The temperature of the water affects the rate of growth of the larvae. Normally metamorphosis is complete after about three months but can take only two months in warm summers; in cool seasons metamorphosis may not be complete by the end of summer and some larvae may over-winter in water.

Reproduction, growth and development

Males seem to defend areas and, under certain conditions, engage in rival combats in which they grasp each other and stand on their hind legs. Apart from the over-wintering period, mating seems to take place at any time. The males search and give chase to females. Then, as part of the mating behaviour, a male carries the female on his back for some time before depositing a spermatophore and the female picks up ensuring this. Gestation takes several months and in most of the sub-species of *S. salamandra* the larvae hatch immediately after the eggs have been produced (the larvae appear in a transparent membrane). Some populations in Spain produce fully formed young; so well developed are the larvae that they resemble miniature adults. The number of young varies widely but is usually about 20. The larvae have prominent feathery external gills, which slowly disappear during metamorphosis. They remain and develop in water up to three months. After metamorphosis, salamanders are strictly terrestrial. Sexual maturity is reached after three to four years.

General comments

Some populations of the European *salamander* have made remarkable adaptations to what would normally be hostile conditions for the species. For example, the relict population on Mount Carmel in Israel has few suitable breeding places, and drying-out ponds in the xeric Mediterranean environment is common. They breed in the early winter, and cannibalism helps survival and metamorphosis of large-sized larvae. Ovoviviparity has been found to occur in sub-species from northern Israel.

Major references

Degani, G. (1981) The adaptation of *Salamandra salamandra* (L.) from different habitats to terrestrial life. *British Journal of Herpetology*, 6, 169-172.

Degani, G. & Warburg, M.R. (1980) The response to substrate moisture of juvenile and adult Salamandra salamandra (L.) (Amphibia; Urodela). *Biology of Behaviour*, 5, 281-290.

Joly, J. (1968) Donnes sur l'ecologiques sur la salamandre tachetee *Salamandra salamandra* (L.) *Annales Sciences Naturelles Zoologiques*, 12e Ser., 10, 301-366.

Warburg, M.R. (1986) Observations on a relic population of *Salamandra salamandra* on Mt. Carmel during eleven years. In Rocek, Z (ed) *Studies in Herpetology*, Charles University, Prague, pp. 389-394.

Warburg, M.R., Degani, G & Warburg, I. (1979) Growth and population structure of Salamandra salamandra (L.) larvae in different limnological conditions. *Hydrobiologica*, 64, 147-155.

Salamandra salamandra. Adult.

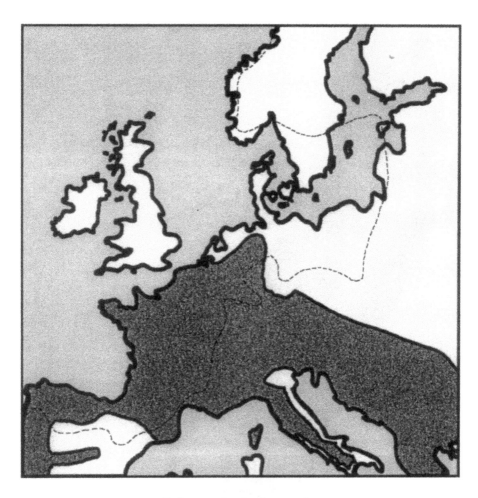

Salamandra salamandra.

Family Salamandridae

Triturus alpestris

Alpine Newt (GB), Triton alpestre (FR), Bergmolch (DE), Alpenwatersalamander (NE), Tritón alpino (ES), Bergvattensalamander (SE).

Introduction

This very beautiful newt was first described in 1768. The name 'Alpine Newt' is a little misleading because, although found in alpine regions, it does occur in some lowland areas and coastal regions as well. It has the curious behaviour of inflating itself and then expelling air with a whistling sound when picked up and handled.

Taxonomy

The Generic name *Triturus* comes from the Latin *tritura* which means 'threshing' and may possibly refer to the characteristic waving and threshing of tails during courtship behaviour. There are at least 8 species of *Triturus* in Europe and 20 distinct forms (though some researchers recognize 12 species). As many as five species may occur in sympatry, that is, share a common distribution. At least five sub-species of *T. alpestris* have been described and some recognize 10 sub-species, though there is doubt about some of these.

Protection

This species is protected in Belgium, Czech Republic, Denmark, France, Germany, Hungary, Luxembourg, The Netherlands, Poland, Spain and Switzerland.

Description

This is a medium-sized newt, variable in size, colour and markings. Females are usually larger than males with a total length about 10 cm (exceptionally 12cm) and less colourful. Sexual dimorphism is less noticeable in this species than in other *Triturus* species. The back of the male is dark grey and the sides are pale with numerous dark spots, bordered by a blue stripe.

There are few or no markings on the orange-red underside. They have a low yellowish dorsal crest, spotted with black, which extends from the neck down the body and most of the tail. After the breeding season, the male loses its crest and the skin of both males and females acquires a slightly warty appearance, which is almost like velvet in the terrestrial phase.

Newly metamorphosed individuals often have a yellow, orange or red dorsal stripe, which may be limited to the region just behind the head. This usually disappears with age but may persist in some adult females.

Distribution and habitat

The *Genus Triturus* is represented in all European countries except Iceland and Malta. The Alpine Newt is not restricted to the Alps and is found in a relatively large area of Europe, as far east as Russia, south to France, northern Spain, Italy and central Greece north to southern Denmark. It occurs at altitudes higher than for any other European Newt and has sometimes been recorded at 1,000 m and occasionally 3,000 m. In Germany there is a positive correlation between altitude and abundance of this species. It has been introduced to Britain and there are now breeding colonies in England and Scotland.

This newt is more aquatic than other European species, sometimes being found in water all year round. It is sometimes referred to as a bottom dweller rather than a free-swimming species, and occupies bodies of water ranging from shallow pools to still water over 2 m deep. Typical breeding sites include freshwater pools, flooded quarries, streams and lakes in mountainous regions. It seems to prefer shaded water in or near woodland; also occurs in small pools in some lowland areas in the northern parts of its distribution. In the southern regions it is strictly a montane species.

Seasonal movements and behaviour

It emerges from over-wintering towards the end of March and remains in water until late August. After leaving the water it spends some time on land before seeking wintering sites in the base of tree stumps, in rotting logs, or beneath stones or boulders.

Vagility and population ecology

When on land, it keeps to shady, moist places. There seem to have been no detailed studies of its movements and population dynamics.

Feeding ecology

The Alpine newt feeds on small crustaceans, adult insects and their larvae, and also worms.

Thermal ecology

Mountainous regions have a wide range of conditions and unstable environments. To survive in alpine regions it seems to need to be resistant to desiccation and able to withstand very cold conditions. It seems able to move about at low temperatures.

Reproduction, growth and development

Mating behaviour takes place in March and April. The male approaches the female, which may then follow the male for a while. They then face each other when the male fans water towards the female. Followed by the female, the male then creeps forward depositing a spermatophore, which is picked up by the female. The eggs (on average 150) are laid singly on submerged vegetation and the larvae hatch in about 15 to 20 days. The larvae typically have large bushy gills and rudimentary forelimbs; the hind limbs develop late. In most populations, metamorphosis takes place after about three months.

There are instances of neoteny, as with other newts, when in some populations the larval form may not metamorphose and yet be capable of reproduction. In one population in Montenegro, for example, there is no metamorphosis at all, and the adults always appear as the neotenous form.

General comments

In a study of genetic variation of three newt species, *Triturus alpestris* was found to have the greatest variation compared to *T. vulgaris* and *T. cristatus*. One theory suggests that there may be a link between environmental variation and genetic variability.

Major references

Arano, B. & Arntzen, J.W. (1987) Genetic differentiation in the Alpine Newt. *Triturus alpestris*. In van Gelder, J.J. et al. (eds) Proceedings of the 4th Ordinary Meeting of the Societas Europaea Herpetologica, Nijmegen, pp. 21-24.

Halliday, R.R. (1977) The Courtship of European Newts: an evolutionary perspective. In Taylor, D.H. & Guttman, S.I (eds) The Reproductive Biology of Amphibians, Plenum Press, New York.

Kalezic, M.L. & Hedgecock, D. (1978) Genetic variation and differentiation of three common European newts. British Journal of Herpetology, 6, 49-57.

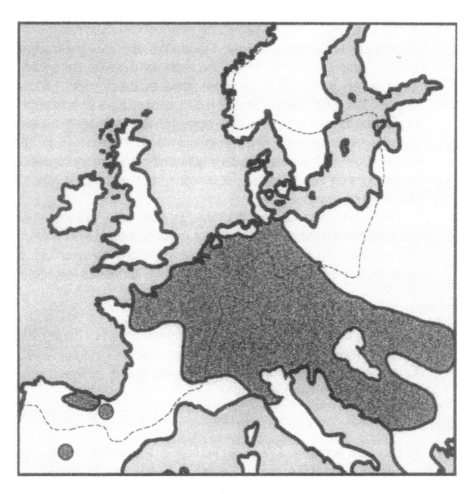

Triturus alpestris.

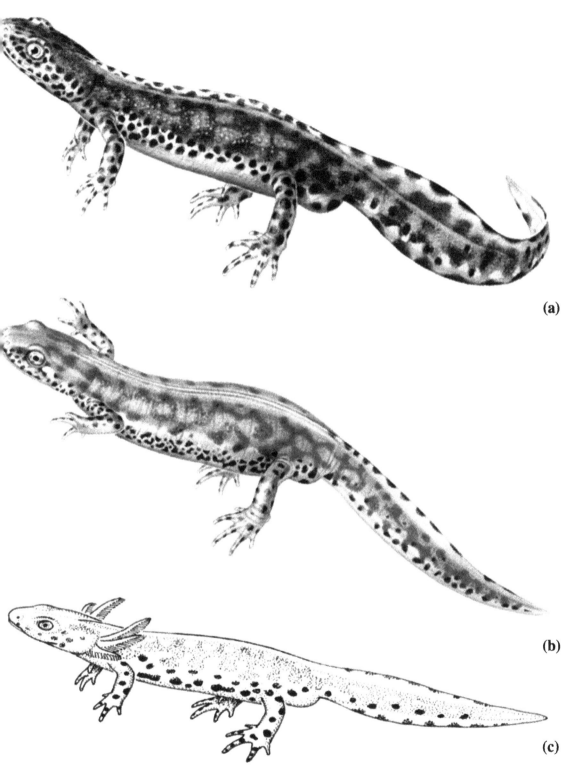

(a)

(b)

(c)

Triturus alpestris. Male (a), female (b) and neotenous individual (c).

Family Salamandridae

Triturus cristatus

Great Crested Newt or Warty Newt (GB), Tritón Crêté (FR), Kammolch (DE), Kamsalamander (NE), Tritón Crestado (ES), Större vattensalamander (SE).

Introduction

The behaviour of newts is best observed at night with the aid of a torch. It is then that the sight of breeding males with their well developed, undulating crests and agile behaviour is not easily forgotten. Their powerful tails are used for swimming and some individuals have been seen to curl their tail around plant stems. The voracious feeding behaviour of the larvae has prompted observations for many years. A certain Mr Higginbottom in 1850 recorded that "I have seen a Warty Triton in its branchial stage, with three of the smaller species in its stomach at one time". The "three smaller species" were presumably larvae or adults of other newt species.

Taxonomy

There are a number of closely related species, some of which were previously recognized as sub-species, e.g., the Northern Crested Newt *Triturus cristatus cristatus*, the Alpine Crested Newt *T.c. carnifex*, the Danube Crested Newt *T.c. dobrigicus* and the Southern Great Crested Newt *T.c. karelinii*. The Alpine species has been reported from some sites in Britain.

Protection

Although sometimes very common locally *Triturus cristatus* has a fragmented distribution and the number of populations has declined dramatically in recent years. For example, in Britain the results of a questionnaire suggested that in 1975 there had been a recent 75% decline in the use of ponds by this species on agricultural land. It is a species which is protected in most European countries. In Britain it is fully protected under the Wildlife and Countryside Act (Variation of Schedules, 1988, 1991). It is also listed in Appendix 11 of the Berne Convention.

Description

The appearance of this newt species is variable. It can sometimes be confused with *Triturus alpestris* or even *Salamandra atra*. The Great Crested Newt is the largest of the European newts and in general adults may reach 16 or 17 cm in length, including the tail, but are more often about 12 to 14 cm. When viewed from above it appears to have a dark brown, coarse or warty skin, which is finely stippled along the lower flanks with white spots. The underside is yellow or orange with bold black and grey markings and spots – in some individuals the underside is completely black with no markings. The breeding males are unmistakable and have a high dorsal, dentated crest separated from the high caudal crest and a whitish or silvery stripe along the side of the tail. Although its appearance does vary, *T. cristatus* has a low level of genetic variation when compared to *T. vulgaris* and *T. alpestris*.

Distribution and habitat

The Great Crested Newt is found throughout most of Europe except in the southwest, where it is replaced by the Marbled Newt *T. marmoratus*. Although it occurs up to altitudes of 2,000 m it is basically a lowland species and is mostly found below 90 m. It is not found in Ireland, southern France, Spain, southern Greece or the Mediterranean islands.

For breeding it tends to favour comparatively deep bodies of water, at least 50 cm to 2 m deep, and it is not unusual to find it in neglected swimming pools. pH and other chemical characteristics of water affect the distribution of newts. The Great Crested Newt, for example, prefers water with a neutral or slightly alkaline pH ranging from about 6.5 to 8.5, and is only rarely found in water with a pH below 6.0 except, apparently, in Norway's acidic conditions especially where there is a lot of aquatic vegetation.

During the autumn and winter is leads a more terrestrial life and is found under stones or beneath moist vegetation where it sometimes over-winters. In some localities both larval and adult forms remain in water for all or part of the winter. The larval form therefore remains aquatic for the normal metamorphosis period in the following season and leaves the water as a large metamorph.

Seasonal movement and behaviour

Between February and March the breeding adults return to water, moving only at night, where they may remain for 2 to 3 months. The males begin the migration and so early in the year ponds often have far more males than females in them. Later the sex ratio is about equal. In Norway, adults have a crepuscular (or early nocturnal) pattern of activity. At the height of the summer, when nights are very short, maximum activity coincides with midnight.

Vagility and population ecology

Although both adults and juveniles do move from pond to pond, this species appears to have a higher tenacity for its breeding ponds than other newt species. In some attempts to relocate the species it has been found that, depending on the state of the ponds, the relocated individuals may attempt to leave the new locations to which they have been introduced. This behaviour has proved to be a problem in those circumstances where it has been desirable to relocate populations, which are under threat as a result of changes in land use. In captivity it has been known to live as long as 20 years.

Feeding ecology

There seems to be a relationship between the size of a newt and the size of its prey; not surprisingly the prey of the Great Crested Newt includes larger items than those taken by smaller newt species. For example, it may take more Asellus (Water Hog Louse) and gastropods and in general feed on benthic or predominantly bottom-living invertebrates. Leeches are commonly taken by this and other species of newts. There are reports of Great Crested Newts preying on full-grown Smooth Newts and even on a Slow Worm.

Thermal ecology

In Poland the adults emerge from over-wintering sites when the ground temperature reaches about 5° C, and then migration to breeding ponds commences when temperatures reach about 7° C. Courtship and egg laying can occur at quite low temperatures but the optimum range seems to be between 15 and 20° C. There are observations of some newts swimming beneath ice.

Reproduction, growth and development

Breeding males will sometimes interact with each other and such encounters may be evidence that they do maintain territories. However when the males mark areas with cloacal secretions, they probably do so to attract females, not to establish territories. The courtship is typical of newt behaviour and starts by the male responding to pheromones (chemical odours) given off by the female. The male keeps just ahead of the female with his tail alternately lashing from side to side then gently quivering. During courtship, the tail's main role is to send vibrations of a particular frequency (different to other species) which transmits pheromones via the current of water towards the female. Females respond by arching their backs and exhibiting characteristic cat-like stances.

Towards the climax of this courtship, the male deposits a spermatophore on the bottom of the pond and induces the female into a position where she picks up the spermatophore in her cloaca. A few days later, the females lay their eggs singly in folded leaves and on stems of water plants – a total of 200 to 300 eggs may be laid. The newt larvae or tadpoles leave the eggs within about three weeks.

The Great Crested Newt is mature when four to five years old (sometimes as young as 3 or even 2 years old). These newts are long-lived and there is a report of 27 years for a captive female. In the wild they may live for 10 or 15 years but probably not as long as 27.

General

The two species *Triturus cristatus* and *T. vulgaris* are found in the same geographical region and it is not unusual for them to occupy the same pond. No evidence of interbreeding has been found. There has been much research on how the two species avoid competing for resources, or how they share resources such as food and space but, nevertheless, there remain many aspects of newt community ecology yet to be investigated.

Major references

Beebee, T.J.C. (1975) Changes in the status of the Great Crested Newt *Triturus cristatus* in the British Isles. British Journal of Herpetology, 5, 481-490.

Dolmen, D. (1983) Diel rhythms of *Triturus vulgaris* (L.) and *T. cristatus* (Laurenti) (Amphibia) in central Norway. Gunnaria, 42, 1-34.

Dolmen, D. & Koksvik, J.I. (1983) Food and feeding habits of *Triturus vulgaris* (L.) and *T. cristatus* (Laurenti) Amphibia) in two bog tarns in central Norway. Amphibia-Reptilia, 4, 17-24.

Glandt, D. (1985) Verhaltungsreaktion und Reproduktion adulter Molche, Gattung *Triturus* (Amphibia, Urodela), nach Langstreckenverfrachtung. Bonner zoologische Beitrage, 36, 69-79.

Green, A.J. (1989) The sexual behaviour of the Great Crested Newt, *Triturus cristatus* (Amphibia; Salamadridae). Ethology, 83, 129-153.

Hagstrom, T. (1979) Studies on the population ecology of the two newt species (Urodela, Triturus) in southwestern Sweden. Dissertation, University of Gothenburg.

Hagstrom, T. (1979) Population ecology of *Triturus cristatus* and *Triturus vulgaris* (Urodela) in SW Sweden. Holarctic Ecology, 2, 108-114.

Latham, D.M., Oldham, R.SW., Stevenson, M.J., Duff, R., Franklin, P. & Head, S.M. (1996). Woodland management and the conservation of the Great Crested Newt (*Triturus cristatus*). Aspects of Applied Biology, 44, 451-459.

Zuiderwijk, A & Sparreboom, M. (1986) territorial behaviour in crested newt *Triturus cristatus* and marbled newt *T. marmoratus* (Amphibia, Urodela). Bijdragen tot de Dierkunde, 56, 205-213.

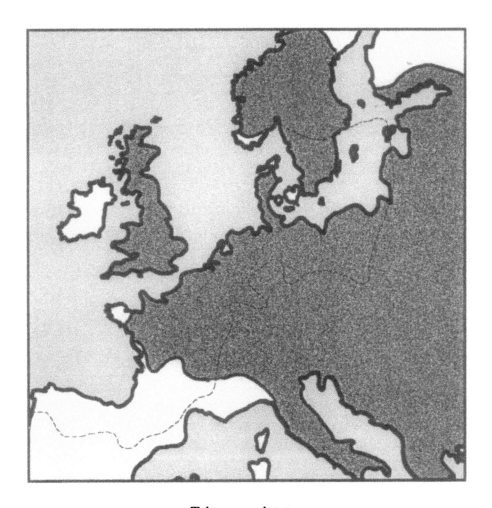

Triturus cristatus.

Triturus cristatus. **Male (a), female (b).**

Family Salamandridae

Triturus helveticus

Palmate Newt (GB), Triton Palmé (FR), Fadenmolch (DE), Vinpootsalamander (NE), Tritón Palmeado (ES), Trådsalamander (SE).

Introduction

This small newt was first described in 1789 by the naturalist Razoumowski. He described specimens collected in Switzerland, thus the specific name helveticus. This newt was originally given the name *Lacerta helvetica*.

Taxonomy

In appearance this species may easily be confused with the larger species *Triturus vulgaris*. Hybrids between *Triturus vulgaris* and *T. helveticus* have been produced in captivity but there is only one authenticated hybrid from the wild. At least four sub-species are recognized and they are morphologically similar to *Triturus montandoni* from the Carpathian Mountains.

Protection

Habitat damage and loss as a result of land use change and intensification has resulted in this newt species disappearing from many areas. It is protected in Belgium, France, Luxembourg, the Netherlands, Spain and Switzerland. In Britain it is listed in Schedule 5 of the Wildlife and Countryside Act and is protected to the extent that it is illegal to sell it.

Description

The females of *Triturus helveticus* are longer than the males and grow to about 8.2 cm including the tail (average is 6 to 7 cm). Average body size has been found to be smaller in the northwestern Iberian Peninsula compared to European populations. Morphometric studies have revealed considerable variation among populations in neighbouring areas. It has a smooth skin (velvet texture when dry) and on the head there are three fairly visible grooves. It is olive green or brown above, sometimes marbled dark green, and yellowish brown beneath with some small spits. The throat is never spotted but has a notable pink tinge. Most females have dark tips to their tails.

The breeding males have ridges of tissue along the back, which result in a box-shaped appearance; the ridge of tissue is similar to, but not comparable with, a crest. The males are pale greenish brown with green and black spots. The ventral side is pale yellow and slightly spotted. The tail ends in a small dark filament (4-7 mm) and the hind feet are black and webbed, often so much that the webs sometimes extend beyond the length of the digits. Outside the breeding season the tail filament, webs and crest are not present. It has been suggested that webbing has evolved in male palmate (and smooth) newts to counteract the potential imbalance caused by relatively bulky tails, especially when swimming down from the water surface. Such a suggestion needs investigation since no appendages are evident in many similar species.

It is very difficult to distinguish between larvae of *Triturus helveticus* and those of *T. vulgaris*. The same is true of new metamorphs.

Distribution and habitat

This newt is found throughout Western Europe from Northern Iberia to Scotland. It is not found in Ireland. Although its distribution seems related to altitude – there is a positive correlation between altitude and abundance– it seems that geological and other features, rather than just altitude, are factors, which determine the distribution.

On land *Triturus helveticus* seems to prefer wooded areas. It breeds in a variety of waters from still to slow running and from large shallow areas to woodland puddles and cart tracks. It is often found in heathland ponds with a low pH, sometimes 4 to 4.5. In some localities this species may occur in brackish coastal ponds. A detailed study of the sub-species *T. helveticus helveticus* in Belgium found that in general it is found in permanent, shaded, clear and medium-large bodies of water. It also seems to avoid polluted or organically enriched water.

Seasonal movements and behaviour

These newts usually over-winter on land but, in common with many other newt species, some may spend the winter in water. On occasions males in breeding condition have been recorded in ponds as late as September. After over-wintering they begin to move to breeding sites during March or April but sometimes as early as February. Many will return to the same breeding ponds year after year.

Population ecology

The sex ratio seems to be slightly in favour of the males. It is not possible to distinguish between different age classes because, beyond the first two years, there is very little growth.

Feeding ecology

The food of this species is the same as that for *Triturus vulgaris*. In captivity it has been found to take frog tadpoles but not Common Toad (*Bufo bufo*) ones. Male Palmate newts with no experience of either frog or toad tadpoles very quickly learn to distinguish between them and reject the toad larvae.

Thermal ecology

From ecological studies on *Triturus helveticus* in France, it seems that the species tends to select cooler ponds than those by *T. vulgaris*.

Reproduction, growth and development

The courtship behaviour is similar to *T. vulgaris* but the use of the tail is more delicate with fanning rather than a vigorous whip action. This fanning movement creates a water current, which almost certainly carries pheromones to the female. The female picks up the spermatophore in the same way and this is followed by internal fertilization. Eggs are placed singly on submerged vegetation and each female lays between 300 and 400 eggs.

On hatching the larvae are only 5-6 mm and larval development takes at least 25 days. The larvae metamorphose in June or July of the following year. A small proportion do not metamorphose and become neotenic adults.

General

It is not uncommon to find this species with *T vulgaris* and it seems that there is a high degree of niché overlap between the two along both microhabitat and seasonal niché dimensions. There is a high degree of overlap with the timing and length of migration and with their food. The ability of these two species to coexist with very similar nichés may depend on unlimited levels of resources.

Major references

De Fonseca, P. & Jocque, H. & R. (1982) The Palmate Newt *Triturus helveticus helveticus* (Raz.) in Flanders (Belgium). Distribution and habitat preferences. Biological Conservation, 23, 297-307.

Galán, P. (1985) Morfologia y Fenologia del Tritón Palmeado, *Triturus helveticus* (Razoumowsky, 1789) en el noroeste de la Peninsula Iberica. Alytes, 3, 31-50.

Van Gelder, J.J. (1973) Ecological observations on amphibia in the Netherlands. II. *Triturus helveticus* Razoumowski: migration, hibernation and neoteny. Netherlands Journal of Zoology, 23, 86-108.

Griffiths, R.A. (1987) Microhabitat and seasonal niche dynamics of Smooth and Palmate Newts, *Triturus vulgaris* and *T. helveticus*, at a pond in mid-Wales. Journal of Animal Ecology, 56, 441-451.

Hagstrom, T. (1979) Studies on the population ecology of the two newt species (Urodela, *Triturus*) in southwestern Sweden. Dissertation, University of Gothenburg.

Reading, C.J. (1990) Palmate Newt predation on Common Frog (*Rana temporaria)* and Common Toad (*Bufo bufo*) tadpoles. Herpetological Journal, 1, 462-465.

Triturus helveticus. **Male (a), female**

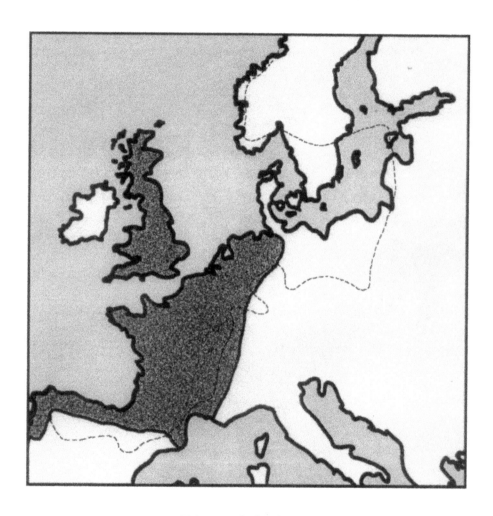

Triturus helveticus.

Family Salamandridae

Triturus vulgaris

Smooth Newt (GB), Triton ponctue ou Triton vulgaire (FR), Teichmolch (DE), Kleine watersalamander (NE), Tritón Europeo (ES), Mindre vattensalamander (SE)>

Introduction

This small newt is probably the most frequently encountered as it is so common and ubiquitous in its distribution. It was first described by Linnaeus in 1758 and was placed in the Genus Lacerta with the European lizards. The specific name simply means common. It appears that Smooth Newts have long been the subject of much mistreatment as this quote from Cooke's book, Our Reptiles and Batrachians (1893) suggest: "these poor creatures are often the subject of persecution in rural districts. Schoolboys, especially, consider them fair game for torture, and adults view its inflection with complacency because not a few still entertain superstitious or fabulous notions either of their poisonous properties, or their secret association with 'black art'."

Taxonomy

The Smooth Newt is also well known because it has been widely studied. At least four sub-species, possibly seven, are recognised and so only a general description can be given here.

Protection

Changes in land use and farming intensification have for many years damaged wetlands, and the subsequent losses of these habitats have led to a very fragmented distribution of the common newt. It is protected in many European countries, for example in Britain it is listed in Schedule 5 of the Wildlife and Countryside Act – it is illegal to sell or offer them for sale. It is fully protected in Northern Ireland.

Description

The sexes are similar in size, and fully-grown adults reach on average 9 cm and occasionally about 11 cm in total body length. In some regions they are much smaller than this. While in water they have a smooth skin; on land their skin loses its glossy appearance and becomes slightly coarse or powdery in appearance. The colour is rather variable but in general they may range from a pale yellow-brown, grey or almost black above, and bright brick red or orange on the underside. During the breeding season, the males have an undulating dorsal crest extending along the back to the tail, and may have 'feathered' toes or delicate lobes on the toes. The crest is most prominent during April to June, and the height of the crest may indicate to females an ability to produce large quantities of sperm. The back and sides are pale greenish brown to grey in colour and the underside is orange. Irregular dark spots over much of the body and parallel blue and red stripes at the base of the tail are characteristic of the breeding males. The females never have the dark spots but may be slightly or even completely speckled. It is these secondary characteristics which are most variable and which have been used as a basis for the identification of sub-species.

Distribution and habitat

This and the Great Crested Newt have a more widespread geographical distribution than any other European species of tailed amphibian. The Smooth Newt is found in almost all parts of Europe (and in many parts of western Asia) except in Southern France, Iberia, southern Italy, and most of the Mediterranean islands.

It is frequently found out of water, under logs and stones in damp areas, and is commonly mistaken for a small lizard. It has even been found over-wintering in seaweed-covered walls near the seashore. It breeds in ponds of all sizes, with a pH not less than 6.5, and may remain there for the summer. Smooth Newts are not uncommonly found coexisting with Great Crested and Palmate Newts. Although it seems to breed in ponds of all sizes and conditions, especially if there is well-developed aquatic vegetation, it does not do so in running water. It is, however, a fairly terrestrial species and occurs quite widely on land during the rest of the year. It is found on some agricultural land, in woodlands, heathlands and gardens.

Seasonal movements and behaviour

Most of the newts to be found in a pond would have migrated there over a period of about four weeks in early spring. Some, however, may have over-wintered in the pond. The amount of time spent in their aquatic habitat can range from about 50 to as many as 200 days. The adults and newly metamorphosed young undergo another migration to land during late summer or autumn. They do not seem to move far and consequently have fairly high site tenacity.

Vagility and population ecology

There seems to be general support for the view that Smooth Newts do have a home range. In Norway, for example, they have been found to move within an area of about 35 square metres. The juveniles seem to be more vagile than the adults. Adults grow at variable rates, sometimes about 1 to 2 mm per year. Some studies have attempted to identify age-classes by measuring newts to within an accuracy of 0.1 mm but size itself may not be a good basis for estimating ages. Other population studies, which have used bone sections, have found that, compared to the development during the first two years of life, there is very little subsequent growth.

Feeding ecology

Smooth Newts like so many other newts, are voracious carnivores and feed on organisms ranging from *Daphnia* to juvenile newts. When food is abundant they take prey of varying sizes but when it is scarce they select whatever gives the best return in terms of effort and size of prey (optimal foraging). The young feed on animals such as chironomid larvae and small gammarids. Smooth newts usually hunt for food during twilight or at night – females are more nocturnal than males.

Thermal ecology

They seem to arrive at breeding ponds when air temperature is between 10 and 15°C. Egg laying commences when the water temperature is 9°C or higher.

Reproduction, growth and development

Smooth Newts are well known for their courtship behaviour, which is characterized by a complex sequence of interactions between the sexes, but it is the male, which is especially active. The ultimate aim of the courtship is for the male to position the female over a spermatophore. To do this the male displays his undulating tail and his bright colours as part of a courtship 'dance'. It is certain that pheromones play an important part in courtship and that the tail is used to thrust water and pheromones from the male to the female. Tactile behaviour also seems to be important because when the female's head touches the male's tail a signal for the deposition of the spermatophore seems to be given.

In some northern localities there are two periods of egg laying: the early period when newts which arrived at the pond during the previous autumn lay their eggs, and then the slightly later time when most others will lay their eggs. Like other pond-dwelling newts, the Smooth Newt lays its eggs singly – each about 2 mm in diameter – wrapping them in leaves of pond vegetation. The number of eggs laid by a single adult can be anything between 100 and 250 but there are reports of higher numbers. The larvae hatch by breaking the surrounding envelope and also by the use of digestive enzymes. The newly hatched young are delicate and very small, about 7 mm long. They cling to vegetation by means of an oral sucker. At first their growth depends on the remains of the egg yolk but very soon teeth develop and external gills appear. When about 14 to 18 mm in length both limbs have appeared.

General

It seems that both adults and newly metamorphosed individuals have a preferred or predictable migration route between the terrestrial and aquatic habitats. The migration route can be a 'corridor' made up of moist ditches and channels, or even damp ruts in the ground. The chemical odours emitted by certain plants, humidity gradients, celestial and magnetic cues have all been put forward to explain newt migration, but more research is needed. The cues used during migration to ponds may be different to those used when leaving.

Major references

Bell, G (1977) The life of the Smooth Newt (*Triturus vulgaris*) after metamorphosis. Ecological Monographs, 47, 279-299.

Dolmen, D. (1981) Local migration, rheotaxis, and philopatry by *Triturus vulgaris* within a locality in Central Norway. British Journal of Herpetology, 6, 151-158.

Giacoma, C. (1988) The ecology and distribution of newts in Italy. Annuario dell' Istituto e Museo de Zoologica dell' Universita di Napoli, 26, 49-84.

Griffiths, R.A. (1987) Microhabitat and seasonal niche dynamics of Smooth and Palmate Newts, *Triturus vulgaris* and *T. helveticus*, at a pond in mid-Wales. Journal of Animal Ecology, 56, 441-451.

Halliday, T.R. (1975) An observational and experimental study of sexual behaviour in the Smooth Newt, *Triturus vulgaris* (Amphibia: Salamandridae). Animal Behaviour, 23, 291-322.

Harrison, J.D., Gittins, S.P. & Slater, F.M. (1984) Morphometric observations of Smooth and Palmate Newts in mid-Wales. British Journal of Herpetology, 6, 410-413.

Marnell, F. (1998). Discriminant analysis of the terrestrial and aquatic habitat determinants of the smooth newt (*Triturus vulgaris*) and the common frog (*Rana temporaria*) in Ireland. Journal of Zoology (London), 244, 1-8.

Schmidtler, J.J. & Schmidtler, J.F. (1983) Verbreitung, Okologie und innerartliche Gliederung von *Triturus vulgaris* in den adriatischen Kustengebieten. Spixana, 6, 229-249.

Verrel, P.A. & Francillon, H. (1986) Body size, age and reproduction in the Smooth Newt, *Triturus vulgaris*. Journal of Zoology, London, 210, 89-100.

Verrel, P.A., Halliday, T.R. & Griffiths, M. (1986) The annual reproductive cycle of the Smooth Newt (*Triturus vulgaris*) in England. Journal of Zoology, London, 210, 101-119.

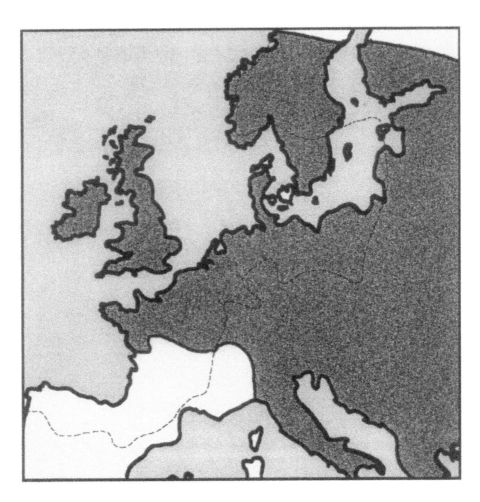

Triturus vulgaris.

Triturus vulgaris. **Male (a), female (b).**

Family Discoglossidae

Bombina variegata

Yellow-bellied or Mountain toad (GB), Sonneur à ventre jaune ou Sonneur à pieds épais (FR), Gelbbauchunke (DE), Geelbuikvuurpad (NE), Sapo de vientre amarillo (ES), Gulbukig klockgroda (SE).

Introduction

The Genus *Bombina*, popularly *known* as the Fire-bellied Toads, is closely related to the Painted Frogs (*Discoglossus*) and Midwife Toads (*Alytes*). These amphibians in the Family Discoglossidae are part of a primitive group, which have small disc-shaped tongues (thus Discoglossidae) which cannot protrude to catch prey. The Yellow-bellied Toad is easily confused with the Fire-bellied Toad *Bombina bombina*, which has a white-spotted underside with black and bright reddish-orange or red markings. They were first described in the mid-eighteenth century and at one stage the Genus was known as Bombinator. Boulenger in 1847 described *B. variegata* as follows: 'this lively little Batrachian, which may be heard in the spring and summer uttering its low and mournful note, hoo, hoo".

Taxonomy

The name *Bombina* is a diminutive of the Latin Bombus, which signifies a booming or deep hollow sound like that of a trumpet, with reference to its calls. This species is known to interbreed with the closely related *B. bombina* found in Eastern Europe. There are narrow clines between populations of the two species. In *B. variegata,* there is considerable variation in morphology and colour, and at least three sub-species have been described: the nominate sub-species, *Bombina variegata variegata*, the Dalmatian *B. v. kolombatovici*, the Italian *B. v. pachypus* and the sub-species from the Balkan peninsula *B. v. scabra*.

Protection

B. variegata is listed in Appendix II of the Berne Convention and is protected in many European counties including Belgium, France, Germany, Hungary, Luxembourg, the Netherlands, Poland and Switzerland.

Description

This is one of the smallest European anurans. It has a short, flattened, characteristic toad body, round head, well-rounded snout and heart-shaped pupils. Males vary in length between 31 and 46 mm and 36 to 50 mm for females. The underside of this small toad is contrasting speckled black and bright yellow or orange. When threatened it arches its back, flexing its head up to display its brightly coloured throat and underside. The dorsal side is basically olive-grey to brownish-grey and some individuals are speckled. Partially albinistic yellow-bellied toads have been found near Frankfurt. There are usually a large number of prominent warts among which are openings of venom glands.

Distribution and habitat

The geographical distribution of this species extends throughout the whole of central and Western Europe, except the Iberian peninsula, but populations occur only sporadically. *B. variegata* has been introduced to Britain and may still survive in southwest England.

This largely diurnal species seems to be able to live in a range of aquatic conditions including temporary pools and ponds, gravel and clay pits, and the edges of rivers and streams. It is not unusual for it to breed in small, temporary bodies of water such as rain puddles, wheel ruts filled with water and roadside ditches. There is a correlation between rainfall and spawning activity in those individuals, which use the temporary bodies of water.

Seasonal behaviour

B. variegata over-winters in soft ground and emerges after the end of April, then migrates considerable distances to breeding sites. At low levels of humidity (54-74%) it has a good sense of direction and is able to locate ponds, but at 90% humidity that sense of direction is lost. When displaced from breeding ponds, most are able to find their way back even over distances as great as 70 m.

Vagility and population ecology

In recapture studies, distances of between 200 and 1200 m between recaptures were recorded: juvenile toads move about far more than adults do. In one study in the Balkan Mountains, males moved about more than females during the spawning season; the distances traveled were much shorter than at the end of the season when several hundred metres may be covered.

In a three year study of this species near Nürnberg, the population was found to consist of three annual age groups; the adult toads (two years), the sub-adults (one year) and younger individuals. One provisional estimate of age structure gave the following: 86% of animals from the first summer, 9% from the second summer, 5% from the third summer or older.

Feeding ecology

Feeding mostly at night, their prey consists of insects (adults and larvae), molluscs and worms.

Thermal ecology

As in other amphibians, temperature affects both vocalization and hearing ability. In *B. variegata*, auditory sensitivity is lacking at 5°C, and then sensitivity greatly increases between 12 and 20°C. The optimum range of temperatures for embryonic development is between 18 and 28°C.

Reproduction, growth and development

This species becomes mature after two winters and the breeding season lasts several months from about mid-April to August. In some localities there may be several distinct, short spawning periods over about three months. It calls in a chorus both day and night but mostly in the evenings. Between May and August the females lay only about 100 eggs in several batches of about 17 each. The size of the eggs varies and this is probably an adaptation to life in a diversity of breeding sites. The species has one of the lowest levels of fecundity for anurans. The larvae, which are about 6mm long, hatch after about 10-12 days but the hatching continues throughout the summer. They have a faster development rate, 5-70 days, than the closely related *B. bombina*, and this has been interpreted as a breeding adaptation for temporary bodies of water. Low temperatures and crowding may result in small toadlets at metamorphosis. The toadlets leave the water any time from late July to early October.

General

The timing, duration and synchrony (or lack) of breeding in amphibians ranges from the 'explosive' strategy of the common frog to the prolonged breeding of *B. variegata*. These differences in breeding ecology and behaviour could be the subject of a useful comparative study of the reproductive strategies of amphibians.

Major references

Barandun, J. (1995) Reproductive ecology of *Bombina variegata*. (Amphibia). Inaugural-Dissertation, Universitas Zürich. 31pp.

Barandun, J. & Reyer, H-U. (1997). Reproductive ecology of *Bombina variegata;* characterisation of spawning ponds. Amphibia-Reptila, 18, 143-154.

Beshkov, V.A. & Jameson, D. (1980) Movement and abundance of the Yellow-bellied Toad *Bombina variegata.* Herpetologica, 36, 365-370.

Kapfberger, D. (1981) Zur Populationsdynamik der Gelbbauchunke (*Bombina variegata variegata* L., 1758). Berichte Naturwissenschaftlich Ges. Bayreuth, 17, 39-45.

Kapfberger, D. (1984) Untersuchungen zu Populationsaufbau, Wachstum und Ortsbeziehungen der Gelbbauchunke, *Bombina variegata* (Linnaeus, 1758). Zoologischer Anzeiger, Jena, 212, 105-116.

Mohneke, R. & Scneider, H. (1979) Effect of temperature upon auditort thresholds in two Anuran species, *Bombina v. variegata* and *Alytes o. obstetricans* (Amphibia, Discoglossidae). Journal of Comparative Physiology, 130, 9-16.

Rafinska, A. (1991) Reproductive biology of the fire-bellied toads, *Bombina bombina* and *B. variegata* (Anura: Discoglossidae): egg size, clutch size and larval period length differences. Biological Journal of the Linnean Society, 43, 197-210.

Seidal, B. (1988) Die Struktur, Dynamik und Fortpflanzungsbiologie einer Gelbbauchunkenpopulation (*Bombina variegata variegata* L. 1758 – Discoglossidae, Anura, Amphibia) in einem Habitat mit temporaren Kleingewassern in Waldviertel (Niederosterreich). Ph.D. thesis, University of Vienna.

Szymura, J.M. & Baron, N.H. (1986) Genetic analysis of hybrid zone between fire-bellied toads, *Bombina bombina* and *B. variegata*, near Cracow in southern Poland. Evolution, 40, 1141-1159

(a)

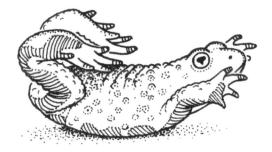

(b)

Bombina variegata. **Adult (a), specimen in a defense postur**

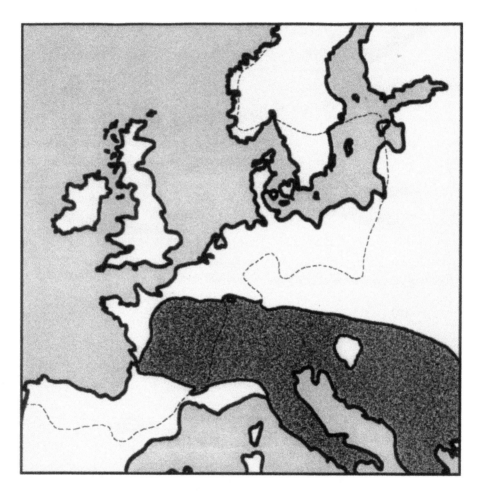

Bombina variegata.

Family Discoglossidae

Alytes obstetricans

Midwife or Bell Toad (GB), Crapaud accoucheur (FR), Geburtshelferkröte (DE), Vroedmeesterpad (NE), Sapo Partero Común (ES), Barnmorskegroda (SE).

Introduction

The Family Discoglossidae is a small, primitive family with seven species in Europe of which three are in the Genus *Alytes*. The name Midwife Toad has been adopted because the male collects the spawn around his legs and carries it around on land until the eggs are ready to hatch. The name Bell Toad is sometimes used because it makes a sound like the chime of a small bell. In 1897 Boulenger described the call as follows: "Towards evening it reveals its presence by a clear whistling note, which has often been compared to the sound of a little bell, or to a chime when produced by numerous individuals".

Taxonomy

There is at least one sub-species in addition to the typical one: *A. o. bascai* from Iberia. There is also an Iberian Midwife toad, *Alytes cisternasii*, and also a very rare, endemic species in Mallorca. *A. muletensis,* which was discovered in 1980. That species is now strictly protected but more provisions are required for the protection if it's habitat.

Protection

The Midwife Toad is listed in Appendix II of the Berne Convention and is protected in many countries such as Belgium, France, Germany, Luxembourg, the Netherlands, Spain and Switzerland.

Description

This small and thickset species grows only up to about 4 to 5cm, and females tend to be slightly larger than males. It is pale grey to pale green dorsally and whitish grey underneath. There are darker patches nearer the shoulders and between the eyes (which have vertical pupils). It has a rounded snout and very short limbs. There are round warts on the dorsal surface and larger ones laterally. Although similar to the Spadefoot Toad and Parsley Frog it can be distinguished from the former by the lack of a 'spade' on the hind foot, and distinguished from the latter by its shorter legs and plumper body.

Distribution and habitat

A western European species, it is found throughout Spain and Portugal and in France except the southeast. It also occurs in Belgium, southern Netherlands, western parts of Germany and northern Switzerland. It was introduced to Britain where it has been in Bedfordshire for around 100 years.

The species is found mainly in hill country and up to altitudes of about 2000m. It inhabits the deserted burrows of other animals and holes in walls but can also create its own burrow using its snout and forelimbs.

Seasonal behaviour

The Midwife Toad does not seem to undertake seasonal migrations. It is a fully nocturnal species and remains concealed during daylight hours. Although it can run like a toad and hop like a frog, its movement is usually slow.

Vagility and population ecology

This species does not appear to be wide-ranging in its movements and individuals tend to remain within a limited area. Not much seems to be known about its longevity and population ecology.

Feeding ecology

Its diet varies according to age and environment (males prefer smaller prey). The variety of prey can be broad: in Spain, for example it consists of arachnids (25%) beetles (22%), collembola (12%), flies (7%) and other insects (16%). Feeding takes place less often during the breeding season.

Thermal ecology

As in other amphibians, temperature effects vocalization behaviour and also hearing. In *Alytes obstetricans* hearing sensitivity is quite well marked even at the low temperature of 5° C, and there is an extraordinary sensitivity up to 20° C primarily in the lower frequencies.

Reproduction, growth and development

The breeding period of this species is nearly the same throughout its whole geographical area. The males attract females by calling at dusk from within their burrows and later from open ground. The call, which is a high-pitched musical note ever few seconds, can be made at temperatures as low as 5 to 7° C. The mating calls are tonal sounds with fundamental frequencies, which increase with a rise in air temperature, and decrease with increased size. That is, the pitch of the call is inversely correlated with male body length.

It is not unusual to see individual females being escorted by several males. Mating takes place on land and eggs are produced also on dry land. While being produced the male fertilizes them and wraps the string of 20 to 50 eggs around his hind legs. About half the males carry eggs from only one female but some carry eggs from two, and a few from three females. The male carries the eggs around his hind limbs for a period of 15 to 45 days (average about 28 days) until the eggs are ready to hatch. At that stage the males move into water where the larvae hatch. Larger males have larger broods and females prefer to spawn with non-brooding males. When females do spawn with already-brooding males they lay smaller clutches. Development of the larvae and metamorphosis takes about a year.

General

The Midwife Toad together with other amphibians, lacertid lizards and the Ascidian Sea Squirt Ciona intestinalis, were used by the Austrian biologist Paul Kammerer (1880-1926) for his studies of acquired characteristics. In 1971 Arthur Koestler published a fascinating and very readable account of Kammerer's life and work. He also gave a sympathetic account of the controversy regarding the inheritance of the appearance of nuptial pads on the male Midwife Toads bred in water. It was the reference to the Lamarckian theory – that acquired characteristics could be inherited – which thrust Kammerer into a storm of controversy.

Major references

Galán, P., Vences, M., Glaw, F., Arias, G.F. & Garcia-Paris, M. (1990) Beobachtungen zur Biologie von Alytes obstericans in Nordwestiberien. Herpetofauna, 12, 17-24.

Koestler, A. (1971) The Case of the Midwife Toad. Hutchinson, London.

Mohneke, R. & Scneider, H. (1979) Effect of temperature upon auditory thresholds in two Anuran species, *Bombina v. variegata* and *Alytes o. obstetricans* (Amphibia, Discoglossidae). Journal of Comparative Physiology, 130, 9-16.

Raxworthy, C.J. (1990) Non-random mating by size in the midwife toad

Alytes obstetricans: Bigger males carry more eggs. Amphibia-Reptilia, 11, 247-252.

Reading, C.J. & Clark, R.T. (1988) Multiple clutches egg mortality and mate choice in the midwife toad, *Alytes obstetricans*. Amphibia-Reptilia, 9, 357-364.

Alytes obstetricans. **Adult.**

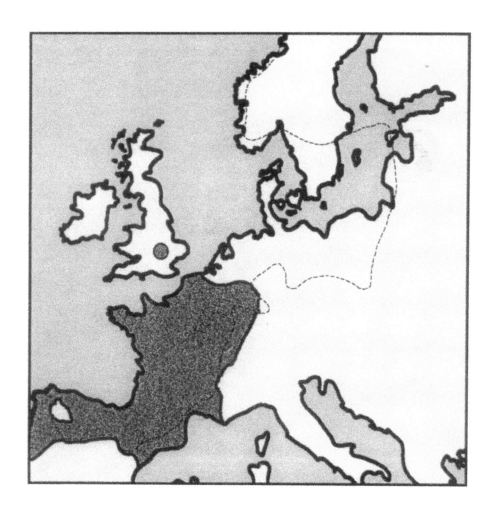

Alytes obstetricans.

Family Pelobatidae

Pelobates fuscus

Common Spadefoot or Garlic Toad (GB), Pélobate brun (FR), Knoblauchkröte (DE), Knoflookpad (NE), Lökgroda (SE), no Spanish name is given because it does not appear on the Iberian peninsula.

Introduction

This is not an uncommon species but possibly because of its nocturnal and burrowing habits, there have been few ecological studies. It is well adapted to burrowing in soft ground and can bury itself to depths up to 1m in a very short space of time. It sometimes emits a garlic smell and consequently is known as the Garlic Toad. When disturbed or threatened by a predator it inflates its lungs, opens its mouth and makes a wailing, screeching sound. This anti-predator behaviour was graphically described in an early work as follows: "When suddenly seized it produces a smell like garlic and utters a startling shrill cry much like that of a kitten, at the same time opening its mouth in a defiant attitude. Prolonged screams can only be compared in a diminutive way to those of an infant."

Taxonomy

The Pelobatidae is a Family of rather primitive amphibians distributed in North African, Southern Asia and North America with only three species found in Europe. The name Pelobates is from the Greek Pelos, earth or mud, and Bateo to walk or tread, hence something like earth treader. The specific name is Latin for brown, tawny or liver-coloured. There is a recognised sub-species *P. f. insubricus* of northern Italy. Other species in the Genus are *Pelobates cultripes* of Iberia and southwest France, *P. varidaldi* of Morocco, and *P. syriacus* of the Balkans and southwest Asia.

Protection

This species is listed in Appendix II of the Berne Convention and is protected in many countries including Belgium, the Czech Republic, Denmark, France, Germany, Hungary, the Netherlands, Poland and Sweden.

Description

> This is a robust, smooth-skinned toad which grows to about 6 to 8 cm in length. Females are larger than males. It has a broad head, which appears arched at the back, with a well-marked dome on top. The dorsal colour varies considerably but is often pale brown in the male and pale grey in the female with olive or brown markings mixed with small reddish spots. The ventral side is pale grey. It has bulging eyes, a slit-like vertical pupil and invisible tympanum. Underneath the hind feet there is a sharp, horny spade-like structure, yellow-brown in colour, which is the 'spade', a modified metatarsal tubercle.

Distribution and habitat

> This species is found over a large area extending from France to southern Sweden and Siberia, extending south to Asia. It is not found in the Iberian peninsula. It seems to prefer low-lying sandy areas and does not occur in mountainous regions.
>
> The habitat is based on soft, sandy soils in woodlands and also in urban areas such as private gardens. This toad is found in water only during the breeding season and seems to prefer rather deep pools. It seems that it may be attracted to eutrophic water.

Seasonal behaviour

> After over-wintering from September to March, movement of large numbers commences rather suddenly. They are a very terrestrial species and have the ability to bury themselves with the help of their powerful 'spades' at a speed that depends on the texture of the soil.

Vagility and population ecology

> *Pelobates fuscus* is nocturnal and seems to have a limited dispersal, becoming particularly active after periods of rain – moving over distances of about 100 m from ponds. It spends short periods in breeding ponds.

Feeding ecology

> Its prey consists mainly of ground beetles and moths but it also feeds on other insects, snails and centipedes. The larvae feed on algae and detritus.

Thermal ecology

> The male usually calls while in water and the incidence of the calls is temperature dependent within the range of 4 to 24° C. High temperatures and also humidity are negatively correlated with the commencement of daily calls.

Reproduction, growth and development

Migration to the spawning ponds takes place in March and April, and males arrive 10 to 20 days before the females. Adults are found in water between about March and May then leave as soon as the eggs are produced. If there is a high level of rainfall during the summer, a second spawning may occur between June and mid-August. The breeding males have a variety of calls and may call when under water. The eggs are laid in gelatinous strings along plant stems, and each female may lay, typically, between 1,200 and 3,400 eggs. The spawn seems particularly vulnerable to attack by moulds. The larvae hatch in four to five days, are relatively large and can grow up to 16cm in length.

Major references

Muller, B. (1984) Bio-akustische und endockrinologische Untersuchungen an der Knoblauchkrote *Pelobates fuscus fuscus* (Laurenti, 1768) (Salientia: Pelobatidae). Salamandra, 20, 121-142.

Stocklein, B. (1986) Untersuchungen and Amphibien – Population am Rande der mittelfrankischen Weiherlandschaft unter besonderer Berucksichtigung der Knoblauchkrote (*Pelobates fuscus* Laur.) Dissertation, Friedrich-Alexander University Erlangen, Nurnberg.

Strijbosch, H. (1979) Habitat selection of amphibians during their aquatic phase. Oikos, 33, 363-372.

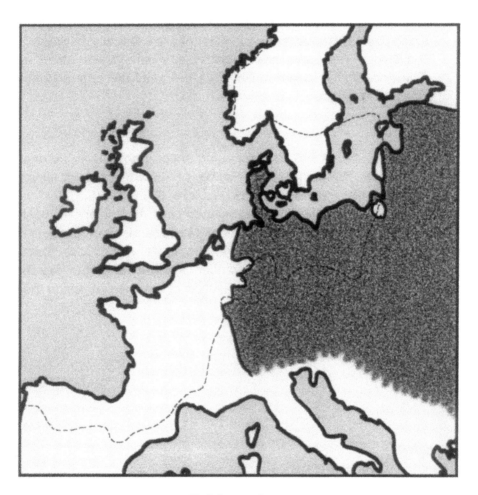

Pelobates fuscus.

Pelobates fuscs. **Adult.**

Family Bufonidae

Bufo bufo

Common Toad (GB) Crapaud commun (FR), Erdkröte (DE), Gewone pad (NE), Sapo Común (ES), Vanlig padda (SE)

Introduction

For some the Common Toad with its jewel-like eyes is a lovely animal, but to many it is about as attractive as a lump of soil. The English essayist and eccentric, Oliver Goldsmith (1730-1774), was obviously not attracted to toads and frogs when he wrote, "If we regard the figure of the toad, there seems nothing in it that should disgust more than that of the frog". He went on to express his horror when he mistakenly dissected a toad instead of a frog. The Common Toad, however, does seem to attract much interest, no more so than when it tries to cross roads in large numbers. There has been so much concern about toad mortalities on roads that toad tunnels and warning signs are in operation throughout Europe. In the Czech Republic, closed roads are policed while toad migrations take place. Warning signs for amphibians were used in Switzerland in the early 1960s, and in Britain the first official toad warning sign was used in 1979.

Taxonomy

The name *bufo* is Latin for toad. This species was originally described by Linnaeus in 1758 and was often referred to as *Bufo vulgaris* after 1768.

Protection

Although widely distributed and very common locally, this species is protected in many European countries. It is protected in Britain by the Wildlife and Countryside Act 1981, which states that it may not be sold or even offered for sale.

Description

The Common Toad is a robust, thickset species with a broad round head and is only superficially like the Common Frog. The skin is dull and wrinkled.

There is a large geographic variation in size and in some localities the females grow up to 15 cm in length. In Britain the females may reach about 11.5 cm in length. Females are larger than males by about 25% in Britain or by 16% in Spain.

Dark-coloured nuptial pads on the inner three digits can distinguish the males. This colour ranges from yellowish-brown to greyish and almost black. Underneath it is white with grey-brown markings. The back is covered with a warty skin and the large bulging leaf-shaped paratoid glands are very clear. Its eyes are golden or copper-coloured.

Distribution and habitat

This widely distributed species is found throughout Europe, temperate Asia and even Japan. In Britain also it has a wide distribution but not in Ireland or the Isle of Man.

It uses a variety of habitats, some of which may be relatively dry. Breeding sites are equally varied and almost any moderate to large body of water seems suitable. In Sweden it has been found breeding in brackish water. It will use the same breeding pond year after year.

Seasonal behaviour

Common Toads over-winter in the ground, in burrows of other animals and even in unheated glasshouses, cellars and outhouses. Every spring, hundreds of thousands migrate from wintering to breeding sites. They tend to travel straight to ponds regardless of the distance. Males arrive first and smaller males tend to arrive before the larger ones.

The time of the start of the annual migration varies from locality to locality and seems to be initiated by both endogenous factors as well as exogenous variables such as day length, humidity, rainfall and temperature. Mortality levels of 45% on roads during the migration have been recorded. Common Toads may travel several hundred metres at an average speed of 30 m per hour on their way to breeding ponds.

Vagility and population ecology

When migrating to breeding ponds Common Toads travel at a rate of about 100 to 250 m per night (when active). During the summer they occupy home ranges of about 1846 square metres at distances between 55 and 1,600 m from the breeding ponds. Although the home ranges may overlap, aggressive behaviour between individuals has not been recorded.

Males commence breeding after two winters and females after four to five years. There are some reports of Common Toads living for 40 years yet there seems to be a large mortality. In one study, for example, on 11% of male toads marked in one year returned to the same pond to breed the following year, and very few were seen after two years.

Feeding ecology

They are opportunistic and take a wide range of live invertebrates, including agricultural pests such as weevils, as well as small reptiles, nestling small mammals and nestling birds.

Thermal ecology

Although temperature and rainfall are factors affecting migration, there is considerable variation in the timing of this activity. In the south of England, for example, they have been known to arrive at ponds as early as mid-January and begin spawning by mid-February. In other localities further north the migration may not take place until March followed by spawning in April. It seems, however, that until they enter the breeding ponds movement is temperature dependent and generally does not take place below air temperatures of 3 to 4° C. Once in the water, temperature seems to have less of an influence.

After spawning, they are not uncommonly found at night feeding on land especially if the temperature is above 10° C. Like the larvae of many other frogs and toads, those of the Common Toad aggregate around the edges of ponds during the day and disperse to deeper water at night. This may be a form of thermoregulatory behaviour.

Reproduction, growth and development

The breeding season is brief and lasts between two and six weeks, depending on temperature. It has long been observed that males seem to outnumber females and that there is much competition amongst males. Several males seeking amplexus with one female is a common sight. Breeding advantage seems to increase with size.

Larger toads have deeper croaks and it seems that male calls can act as a signal in antagonistic encounters between males. Female toads may even be able to distinguish which are larger males by the pitch of their calls. Fecundity is related to size and larger females produce more eggs. Spawning takes place in late March and early April but may commence much earlier if the weather is mild. Fully-grown females can produce up to 6,000 eggs in strings up to 2m long. The larvae begin to emerge after a few weeks and metamorphose after three months with more rapid development occurring at higher temperatures.

General

There have been few long-term ecological studies on amphibians in Europe. One of the few examples is the work on the Common Toad, which commenced in the south of England in 1979. It is this kind of research that is invaluable for understanding the population ecology of amphibians.

Major references

Banks, B. & Beebee, T.J.C. (1986) A comparison of the fecundities of two species of toad (*Bufo bufo* and *Bufo calamita*) from different habitat types in Britain. Journal of Zoology, London, 2008, 325-337.

Davies, N.B. & Halliday, T.R. (1978) Deep croaks and fighting assessment in toads *Bufo bufo*. Nature, 274, 683-685.

Gittens, S.P., Parker, A.G. & Slater, F.M. (1980) Population characteristics of the Common Toad (*Bufo bufo*) visiting a breeding site in mid-Wales. Journal of Animal Ecology, 49, 161-173.

Heusser, H. (1968) Die Lebensweise der Erdkröte *Bufo bufo* (L.) Wanderungen und Sommerquartiere. Revue Suisse Zoologie, 75, 927-982.

Reading, C.J. (1998). The effect of winter temperatures on the timing of the breeding activity in the common toad *Bufo bufo*. Oecologia, 117, 469-475.

Reading, C.J. & Clarke, R.T. (1983) Male breeding behaviour and mate acquisition in the Common Toad, *Bufo bufo*. Journal of Zoology, 201, 237-246.

Reading, C.J., Lowman, J. & Madsen, T. (1991) Breeding pond fidelity in the common toad, *Bufo bufo*. Journal of Zoology, 225, 201-211.

Sinsch, V. (1987) Migratory behaviour of the toad *Bufo bufo* within its home range and after displacement. In van Gelder, J.J. et al., eds. Proceedings of the Ordinary General Meeting of the Societas Europaea Herpetologica, Nijmegen. Pp. 361-367.

Sinsch, V. (1988) Seasonal changes in the migratory behaviour of the toad *Bufo bufo*: direction and magnitude of movements. Oecologia, 76, 390-398.

Bufo bufo. **Adult.**

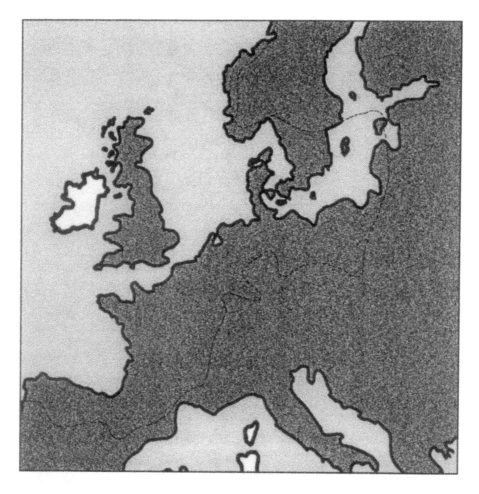

Bufo bufo.

Family Bufonidae

Bufo calamita

Natterjack Toad or Running Toad (GB), Crapaud calamite ou Crapaud des joncs (FR), Kreuzkrote (DE), Rugstreeppad (NE), Sapo Corredor (ES), Stinkpadda (SE).

Introduction

> The yellow vertebral stripe (some vernacular names refer to this) and the means of locomotion of this species are distinctive – Natterjacks crawl rather than hop – have long been the subject of comment as demonstrated by this old quote reported in Cooke's 1893 book, *Our Reptiles and Batrachians:* "…in its pace it differs from the rest of the toad tribe, running nearly in the manner of a mouse".

Taxonomy

> The name *Bufo calamita* means Reed Toad. Although closely related to the other European bufonids *Bufo bufo* and *B. viridis*, the amount of hybridization is limited. This toad is geographically widespread and although clines may exist, detailed studies of morphology and blood proteins of different populations throughout Europe have not provided sufficient evidence for the existence of sub-species.

Protection

> The extensive decline in populations and habitats of this species in several countries has been rapid. In Britain, for example, there was a reported overall decline of 75 to 80% between 1940 and 1981. The major cause of that decline has been the loss of habitat, especially the lowland heaths of southern and eastern England to agriculture, forestry and housing. Concern for the decline of this species has led to much ecological research, management programmes, habitat restoration and of course drafting of legislation. This species now has wide legal protection: it is listed, for example, in Schedule 5 of the UK Wildlife and Countryside Act 1981 and it is in Appendix II of the Berne Convention. It is therefore fully protected in most European countries.

Description

Generally marbled with olive-green or brown, the yellow stripe along the back is unmistakable – very few individuals, except some from Spain, have not stripe. There is little difference in size between sexes and they grow to about 8 cm in length. The hind legs are short and not much longer than the front limbs. The males have a striking large vocal sac beneath the skin, which is particularly noticeable during breeding.

Distribution and habitat

Although this species is well adapted to warm, arid habitats it has spread successfully through much of Europe. It is found in the greater part of Western Europe, the Iberian peninsula, and east to the Baltic countries. It occurs in the south as far as France and Spain, and in the north is found in Britain, Denmark and southern Sweden.

The terrestrial habitat includes heaths, bogs, sand dunes, clay and gravel pits, sandy fields and orchards but not woodlands. *B. calamita* prefers open ground with short vegetation in which it can easily move about and feed. Natterjack Toads occur on open sandy habitats in the Netherlands and dune and heathland habitats in Britain. They typically breed in shallow sandy pools with pH between 6 and 9. In Denmark and Sweden in particular they breed successfully in brackish pools up to 3.5% salinity.

Seasonal behaviour

Not much is known about the wintering sites of this species but it has been found at depths of 30 to 50 cm (sometimes 300 cm) in damp sand during the winter. In most localities, the Natterjack emerges from wintering sites later than its close relative, the Common Toad, and this may be any time between March and June. In the north of Spain, however, it is active earlier than the Common Toad. The males are the first to migrate from over-wintering to breeding sites but they do not migrate in large groups. The females arrive over a protracted period of time and this may be a way of ensuring that at least some spawn is produced at the best time to benefit from temperature and other factors. Females leave the breeding sites immediately after spawning and both females and males disperse throughout May and June. Natterjacks are excellent burrowers.

Vagility and population ecology

There are a few long-term ecological studies of this species from the Netherlands and Norway, and one other study which began in England in 1972. Individual toads tend to keep within an area of about 100 to 200 metres diameter for a short period of time and then may move to another location. It is not unusual for them to abandon a burrow and establish a new area of activity on several occasions during the summer. Compared to similar toad species, populations tend to be rather small.

Natterjacks tend to develop quickly and females may commence breeding when two years old, although three to four years seems more common. They may live to 12 or 16 years and in captivity for longer than that. In a recent study it was found that in high population densities they tend to grow more slowly and erratically, are in poorer condition and forage more widely compared to individuals in low-density populations.

Feeding ecology

Natterjack larvae feed mainly on algae and also on microorganisms. Adults actively forage throughout varied habitats after dusk. They pursue prey with considerable agility over short distances. Detailed studies of types of prey show that more northern populations take a broader range of prey than southern ones. Beetles, flies (mainly crane fly larvae) and hymenoptera are the most common types of prey but Natterjack Toads also take worms, molluscs, spiders, myriopods, true bugs (Hemiptera) and crustaceans. Emerging Caddis flies (Trichoptera) during May are of particular seasonal importance and may contribute to about half of the food intake at that time of year.

Thermal ecology

Studies have shown that this species survives best in relatively warm places. Migration takes place at night, for example, when the temperature has reached about 8° C. Although breeding occurs mostly within the temperature range of 14 to 25° C, weather conditions may affect breeding in quite subtle ways: the temperature of the water during late evening may determine whether calling will occur and for how long. The mean voluntary temperature is about 30° C, and the optimal temperature for larval development is between 20 and 25° C.

Reproduction, growth and development

In the southern regions of its distribution the Natterjack Toad mates from February through to September. In other regions it mates only in the spring and early summer. The breeding season is prolonged, compared to other European anurans, and lacks synchrony, that is it is not an 'explosive' breeder as is the case for the Common toad and Common Frog. The reproductive period may be divided into early, middle and late breeding periods. Females remain in the ponds for only as long as it takes to spawn. Spawning occurs typically in shallow water up to 15 cm deep, and fully-grown females lay between 3,000 and 7,000 eggs in long strings. After about five to eight days the first larvae emerge. The larval stage is brief and shorter than for most other European amphibians – metamorphosis can be complete within six to eight weeks, exceptionally four. Both temperature and tadpole population density (extent of crowding) may affect larval development and survival.

Annual mortality in many amphibians is as much as 95% and the Natterjack Toad is no exception. Both the spawn and tadpoles may be subject to high levels of mortality. One study found that only 0.3% of larvae completed metamorphosis. Although some fish species may have contributed greatly to that high rate, pollution and acidification of breeding ponds, infestation of spawn by fungi, ponds drying out and predation all seem to contribute to high levels of mortality under varying conditions. Furthermore, Common Frog and Common Toad tadpoles can, if there is wide niché overlap, contribute to high levels of spawn destruction.

General

Ecological competition with other amphibians may be an important factor in the survival of this species. In natural pools, for example, tadpoles of *Rana temporaria* can inhibit the growth of those of *Bufo calamita*, even when the two species are separated. A similar effect has been found in artificial ponds.

Major references

Banks, B. & Beebee, T.J.C. (1988) Reproductive success of Natterjack Toads *Bufo calamita* in two contrasting habitats. Journal of Animal Ecology, 57, 475-492.

Beebee, T.J.C. (1979) A review of scientific information pertaining to the Natterjack Toad *Bufo calamita* throughout its geographical range. Biological Conservation, 16, 107-134.

Beebee, T.J.C. (1983) The Natterjack Toad. Oxford University Press, Oxford.

Beebee, T.J.C. et al. (1990) Decline of the natterjack toad *Bufo calamita* in Britain: Palaeoecological, documentary and experimental evidence for breeding site acidification. Biological Conservation, 53, 1-20.

Denton, J.S. & Beebee, T.J.C. (1993) Density-related features of natterjack toad (*Bufo calamita*) populations in Britain. Journal of Zoology, London, 229, 105-119.

Flindt, R. & Hemmer, H. (1968) Beobachtungen zur Dynamik einer Populationer von *Bufo bufo* und *Bufo calamita*. Zool. Jb. Syst., 95, 162-186

Griffiths, R.A. (1991) Competition between common frog, *Rana temporaria*, and natterjack toad, *Bufo calamita*, tadpoles: the effect of competitor density and interaction level on tadpole development. Oikos, 61, 187-196.

Sinsch, V. (1988) Temporal spacing of breeding activity in the natterjack toad. *Bufo calamita*. Oecologia, 76, 399-407.

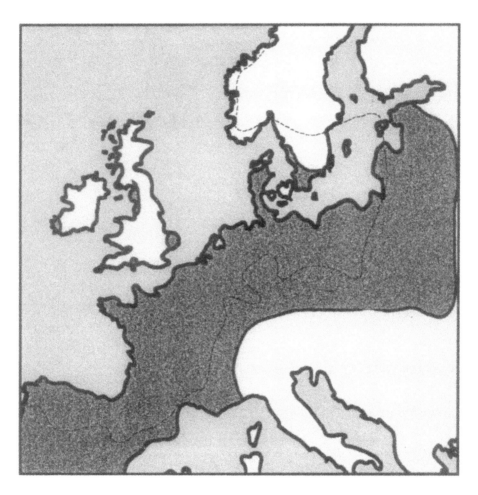

Bufo calamita.

Bufo calamita. **Adult.**

Family Hylidae

Hyla arborea

Common, European or Green Treefrog (GB), *Rainette verte* (FR), Laubfrosch (DE), Boomkikker (NE), Ranita de San Antonio (ES), Lövgroda (SE).

Introduction

The tree frog is well known in folklore and it has often been said that it calls when there is to be a change of weather. It is also said in some old texts that the barometric qualities of this species were so valued in Germany that tree frogs were kept in glass cylinders or terraria furnished with ladders standing in a few centimetres of water. The frog was supposed to ascend or descend the ladder and thus forecast the weather. The powers of this frog have been much exaggerated and although its behaviour is certainly temperature dependent, as is the case with all other temperate zone amphibians, there is no demonstrated ability to foretell the weather.

Taxonomy

There are many kinds of tree frog found throughout the world but they are most abundant in Australasia and the Americas. Most tree frogs are very small species but there are exceptions, for example the Cuban Tree Frog is 13 cm long. One group, the hylids, although absent from most of Africa and Southeast Asia is fairly successful. The East Asian populations once assigned to *Hyla arborea* are now put in other species groups. The tree frogs in northwest Africa are *Hyla meridionalis*, and those on Corsica and Sardinia are often separated as a third European species *Hyla sarda*. Some sub-species are recognised.

Protection

It is protected in most European countries and is listed in Appendix II of the Berne Convention. It is particularly endangered in some regions such as the Netherlands, where it is at the most northwestern part of its distribution, and in northwest Italy.

Description

This is one of the smallest of the European frogs and grows only to between 3 and 5 cm. It has an attractive face and a slender, oval body with a broad head and a blunt snout. Deeply indented webs join its digits and all the tips have small, round, adhesive disc-like pads. The skin is smooth on the dorsal side but underneath it is rough or slightly granular. It is usually a bright leaf-green but this colour can be extremely variable and range from green to dark brown or even yellow. It has the ability to change colour quite rapidly and this seems to be partly temperature dependent and also affected by the animal's physiology. There is a black stripe, which extends from the nostrils down the side of the body to the groin with a short upward branch from the hip. The toes and flanks may have a pinkish colouration. The males have a large yellowish vocal sac beneath the chin. This species is easily confused with the Stripeless Treefrog *Hyla meridionalis* found in the Mediterranean area.

Distribution and habitat

The geographical distribution of this species extends across the whole of continental Europe and Russia, south to Spain and Italy, north to southern Sweden. It is a very rare species in the Netherlands and it has been introduced to at least one locality in England.

The Common Tree Frog is highly selective of habitats. It inhabits varied ponds and pools, but especially those with well developed vegetation. It seems that shore vegetation is important; also shrubs and tussock grasses seem commonly associated with breeding ponds. This species lives in water only during egg laying and for the rest of the time it inhabits vegetation near pools and amp areas. It is mainly nocturnal.

Seasonal behaviour

It over-winters until about March and males are heard to call from vegetation around ponds in April through to about July. Females arrive at the breeding pond after the males.

Vagility and population ecology

Tree Frogs can be both sedentary or migrate over large distances. In one study they moved between pools about 4km apart. Studies on populations in The Netherlands have found 4-5 year old individuals. Body length is not a good indicator of the frog's age.

Feeding ecology

They feed mainly on insects and spiders. One study in Spain revealed that ants were commonly taken, and in another study on the summer ecology of this species in Germany, remains of prey in faecal pellets was composed of 47% Diptera and 34% Coleoptera, many of which were from the Family Cerambycidae, longhorn beetles.

Thermal ecology

Reproductive behaviour seems to be affected by seasonal temperature changes. Similarly, temperature and the onset of darkness have been found to influence the chorus activity – they do not begin to call before evening – and rain and possibly atmospheric pressure may also be influences.

Reproduction, growth and development. They are very vocal and the call is loud, sharp and rapid. The sound of the chorus can carry long distances at night, sometimes as far as 1km. When fully inflated the vocal sac may be almost as large as the head. Spawning takes place during May and June. The small clumps of spawn, only a few centimetres in diameter, containing several hundred eggs are attached to emergent vegetation about 8-10 cm below the water surface. The larvae tend to keep to the bottom and consequently are difficult to find.

General

There have been few ecological studies of the European Tree Frog and a comparative study with the Stripeless Treefrog *Hyla meridionalis* would be useful.

Major references

Clausnitzer, H.J. (1986) Zur Okologie und Ernahrung des Laubfrosches *Hyla a. arborea* (Linnaeus, 1758) im Sommerlebensraum. Salamandra, 22, 162-172.

Pavignano, I., Giacoma, C. & Castellano, S. (1990) A multivariate analysis of amphibian determinants in northwestern Italy. Amphibia-Reptilia, 11, 311-324.

Schneider, H. (1977) Acoustic behaviour and physiology of vocalization in the European Tree Frog, *Hyla arborea* (L). In Taylor, D.H. & Gutman, S.I., eds, the Reproductive Biology of Amphibians, pp.295-335. Plenum Press New York.

Stumpel, H.P. & Hankekamp, G. (1986) Habitat and ecology of *Hyla arborea* in the Netherlands. In Rocek, Z., Studies in Herpetology, pp409-412. Charles University, Prague.

Vos, C.C. & Stumpel, H.P (1995). Comparison of habitat – isolation parameters in relation to fragmented distribution patterns in the frog *Hyla arborea*. Landscape Ecology 11, 203-214.

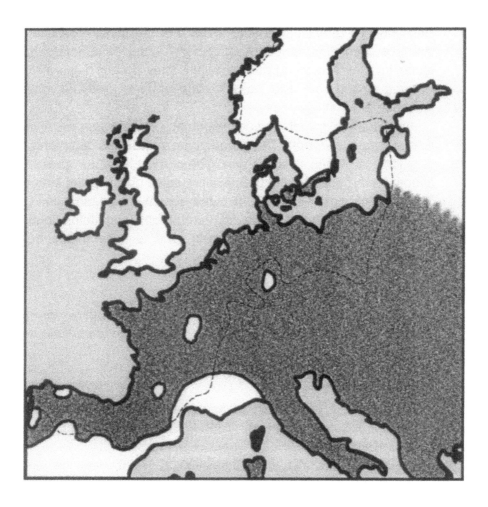

Hyla arborea.

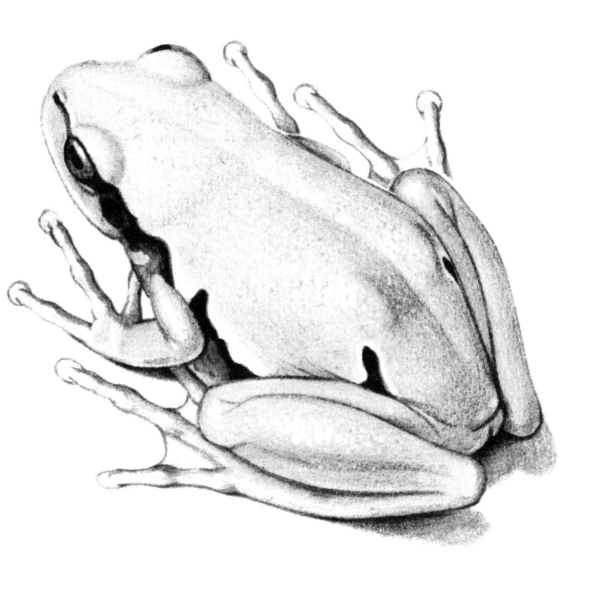

Hyla arborea. **Adult.**

Family Ranidae

Rana arvalis

Moor Frog or Field Frog (GB), Grenouille oxyrhine (FR), Moorfrosch (DE), Heikikker (NE), Rana Campestre (ES), Åkergroda (SE).

Introduction

Suitable habitats for this species have gradually become more and more scattered as a result of land drainage, changes in land use and pollution, especially acidification of breeding pools. Consequently the distribution maps for *R. arvalis* have had a very fragmented appearance. In the nineteenth century this fragmented distribution led to a vigorous debate whether this frog was a boreal species – that is, a species previously of northern Europe and now restricted to cold wet areas in post-glacial times. The zoologist Boulenger (1858-1937) had rejected the idea that *R. arvalis* was a boreal species and noted that "the fragmented and almost spot-like character of its distribution towards the west, and the nature of the localities in which it usually occurs, clearly point to it being in the process of gradual extinction, a relic of a former period, the conditions of which are fast disappearing through human agency".

Taxonomy

This species of 'Brown Frog' shows a preference for wetter habitats than the closely related species *R. temporaria*, and thus the name of Moor or Field Frog. In the southern region of its distribution, it tends to be larger and have longer hind legs; there the sub-species *R. a. wolterstorfii* has been recognised. The specific name comes from the Latin arvalis meaning 'in the fields'.

Protection

There is some concern about the decline in Moor frogs which, in some localities, may be attributed to acidification of the water. It is the rarest brown frog in central Europe and it is listed in Appendix II of the Berne Convention. It is protected in several countries including Belgium, Denmark, France, Germany, Hungary and Norway.

Description

As one of the brown frogs, the colour varies considerably, ranging from reddish-brown with dark spots to pale brown or pale yellow, sometimes with small black spots. There is often a striped dorsal pattern. The underside is yellowish white. Breeding males may have a bluish colouration. Females are larger than males and grow up to 8 cm in length. On average the species is 5 to 6 cm in length with a slender body and a head as long as it is broad. The snout protrudes and distinguishes the species from *R. temporaria*. The diameter of the tympanic membrane is less than the diameter of the eyes. It is very similar to the Common Frog *R. temporaria* except for its more pointed snout and larger and harder metatarsal tubicle.

Distribution and habitat

The geographical distribution extends across central Europe from France to Lake Baikal in Russia, and from northern Sweden to the northern Balkans. It is found in low-lying country, and in hilly regions not usually above 100 m. The habitat consists mainly of damp meadows, marshy fields, fens, sphagnum bogs and moors. Small ponds and ditches are used as spawning sites. It will sometimes perch a few centimetres above ground on the grass.

Seasonal behaviour

R. arvalis over-winters in mud either under water or on the ground, and emerges at the beginning of March, usually slightly later than *R. temporaria*. Juveniles may be active most of the day, adults at night. When on land the species behaves in a curious way to potential attack: a long, high jump is followed by a series of rapid scrabbling, burrowing movements.

Vagility and population ecology

In one study, population density estimates from capture-recapture studies gave values of between 140 and 700 individuals per hectare. During the summer, the home range was found to be 260 square metres.

Moor Frogs are sexually mature when about three years old – about 4 cm in length – but can start breeding at two years. Annual mortality at 40% of the population was found in one study in Sweden. Few survive more than six years.

Feeding ecology

Like Common Frogs, Moor Frogs feed on a wide range of invertebrates including insects and worms.

Thermal ecology

Mating and spawning takes place at temperatures between 4 and 10° C. In one telemetric study, frogs were found to be generally warmer than their surroundings. The body temperature data collected were less variable than expected, suggesting that some behavioural thermoregulation does occur.

Reproduction, growth and development

Mating and spawning are restricted to a very short period of time during the spring. The call of the male has been described as similar to the sound of air bubbles escaping from a bottle under water. The spawn is produced in shallow water, where there is abundant vegetation, towards the end of March and in early April. Females lay between 1,000 and 2,000 relatively small eggs, which sink to the bottom where they develop. High proportions of spawn clumps become infected with fungus and the eggs perish. The larvae metamorphose after about 13 weeks.

General

This is a much threatened but well studied species. In several countries, the results of many ecological research programmes on the species are being put to good use in the design of conservation strategies. Those strategies could usefully be part of long-term monitoring programmes.

Major references

Bellemakers, M.J.S. & Dam, H. van (1992) Improvements of breeding success of the Moor Frog (*Rana arvalis*) by liming of acid moorland pools and the consequences if liming for water chemistry and diatoms. Environmental Pollution, 78, 165-171.

Gelder, J.J. van & Domen, H.C.J. (1970) Ecological observations on amphobia in the Netherlands. 1. *Rana arvalis* Nilsson: Reproduction, growth, migration and population fluctuations. Netherlands Journal of Zoology, 20, 238-252.

Glandt, D. & Podloucky, R. (1987) Der Moorfrosch. Metelener Artenschutzsymposium. Beiheft 19, pp. 161. Biologische Institut Metelen, Niedersachsen.

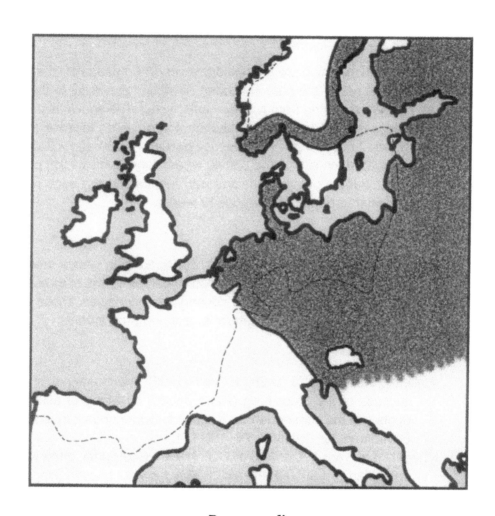

Rana arvalis.

Rana arvalis. **Adult.**

Family Ranidae

Rana dalmatina

Agile Frog (BG), Grenouille agile (FR), Springfrosch (DE), Springkikker (NE), Rana Agil (ES), Långbensgroda (SE).

Introduction

This rather elegant species was first described in 1840 and was once called *Rana agilis*. The name 'agile' is very apt because its extraordinary long, slim hind legs enables it to leap some considerable distances – 2 m have been recorded. Perhaps because of the species' agility it has been described in the past as a popular pet, but only if provided with a large vivarium.

Taxonomy

This is a 'brown frog' with no described sub-species. The name dalmatina was given to a large specimen – 9 cm from snout to vent – from Dalmatia in the Balkans.

Protection

It is listed in Appendix II of the Berne Convention and is protected in many countries such as Belgium, Denmark, France, Germany, Hungary, Luxembourg, Spain and Switzerland.

Description

The females are larger than the males and grow to between 5 and 7 cm. This species has a slender body and a flat head, which is as long as it is broad. The snout is pointed and protrudes in front of the mouth. The colour resembles dead leaves: plain light brown or yellow-buff with a few or no speckles on the dorsal side and yellowish-white without markings on the ventral side. There is a dark temporal spot just behind the eye and there may be a faint vertebral stripe. The large typanic membrane, which is close to, and is the same size as, the eye is a characteristic feature. The forelimbs are relatively short but the hind legs are strikingly long – when stretched forward along the body the feels reach beyond the tip of the head. The groin is yellow. This species is very similar in appearance to *Rana arvalis* and *R. temporaria*.

Distribution and habitat

Found throughout central and southern Europe, including northern Spain and eastwards to Asia Minor and the Caucasus, this frog nevertheless has a particularly fragmented distribution. It is found in the coastal areas of the Danish island (there are some isolated populations on some of the larger ones), southeastern Sweden and also on Jersey (Channel Islands) where it is a particularly threatened species. It has been suggested that the distribution of this species in southern Scandinavia is a relict distribution from the early post-glacial period (8,000 years BP).

The habitat of *R. dalmatina* occurs from sea level to altitudes of about 1,200m, and consists of damp, shady areas near and in woodlands and riverside meadows. It seems to prefer small pools and in one habitat analysis in northwest Italy it was found that its occurrence and population size depended only on the limited extent of deciduous woodlands.

Seasonal behaviour

It seems to have a short wintering period and appears in January or February. The vocalization of the males is similar to that of the tree frogs *Hyla arborea* but is much weaker because it has no vocal sac.

Vagility and population ecology

The males are sexually mature in their second year while females seem not to breed until their third year. These frogs may travel up to about 2 km from water and are crepuscular being particularly active during dusk and at night.

Feeding ecology

They feed on small invertebrates including beetles, flies, Dermaptera and spiders; also molluscs and worms.

Thermal ecology

They seem to prefer warm habitats such as along the edges of woodland on south-facing slopes. The distribution is therefore very fragmented. The development rate of the larvae is temperature dependent and in comparisons with *Rana temporaria* in southern Denmark, it was found that their development time was nearly twice as long as for *R. temporaria*.

Reproduction, growth and development

Mating commences in March and April except in the south were it is in February. The frogs enter the water only to breed and females may lay 700 to 1,000 eggs in one mass. In southern Denmark the first eggs appear in early April and hatching takes place in early May. The larvae metamorphose after about four months. Temperature seems to affect rate of growth but crowding may inhibit or reduce growth of tadpoles. Survival after hatching is exponential and in one study in southern Denmark only 1.8% of eggs had survived after 70 days. The flatworm *Polycellis nigra* is an important predator of the eggs.

General

The levels and rate of mortality of the eggs and larvae must affect the population structure. It has been suggested therefore that the larval stage is possibly very important in population regulation. There has not been much research on the implications of egg and larval mortality on the population dynamics of frogs.

Major references

Ahlen, I. (1984) Theories about the distribution of the agile frog *Rana dalmatina* in Sweden. Proceedings of the 2nd Nordic Symposium on Herpetology. Gunneria, 46, 7-9.

Pavignano, I., Giacoma, C. & Castellano, S. (1990) A multivariate analysis of amphibian habitat determinants in northwestern Italy. Amphibia-Reptilia, 11, 311-324.

Polushina, N.A. & Kushniruk, V.A. (1963) Ob ekologii prytkol lyagushki (*Rana dalmatina* Bonaparte). Zoologicheskii Zhurnal, Moskva, 42, 1881-1884.

Riis, N. (1984) Autecological investigation of *Rana dalmatina* on Funen. Proceedings of the 2nd Nordic Symposium on Herpetology. Gunneria, 46, 22-23.

Riis, N. (1991) A field study of survival, growth, biomass and temperature dependence of *Rana dalmatina* and *Rana temporaria* larvae. Amphibia-Reptilia, 12, 229-243.

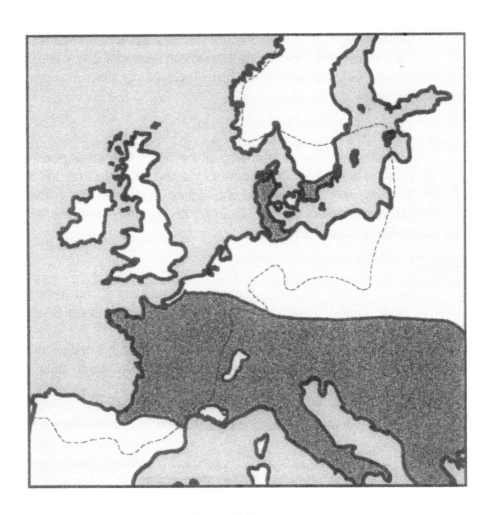

Rana dalmatina.

Rana dalmatina. **Adult.**

Family Ranidae

Rana lessonae

Pool Frog (GB), Petite Grenouille Verte (FR), Kleiner Wasserfrosch (DE), Groene kikker (NE), Dammgroda (Gölgroda) (SE).

Rana esculenta

Edible frog (GB), Grenouille Verte (FR), Teichfrosch (DE), Groene kikker (NE), Rana Europea Común (ES), Ätlig Groda (SE).

Introduction

The name *lessonae* is probably after an Italian called Lessona. The specific name esculenta means edible; the animal itself is the main source of the infamous frogs' legs. The practice of catching and preparing frogs for eating is cruel and quite repugnant. Typically the frogs are not killed prior to processing. A sharp blade is used to sever the rear portion of the animal, leaving the wriggling head and body to die slowly. This is not the only frog species considered edible and many other species are eaten in various parts of Europe and elsewhere. The following quote from 1866 demonstrates even then that these frogs were grossly exploited: "Edible frogs are brought from the country, thirty or forty thousand at a time, to Vienna and sold to the great dealers, who have conservatories for them, which are large holes, four or five feet deep."

Taxonomy

These two species are typical examples of 'green' frog and of all the species described in this book, they have presented the most difficult taxonomic and systematic problems. Many statements about their status as species have been made and sub-species have been suggested on the basis of studies of vocalization, immunology, anatomy and morphology. There seems no doubt that *R. lessonae* is a true species and that *R. esculenta* is hybrido-genetic or klepton (from Greek for 'thief', because it needs a chromosome from another species to survive – Gunther, 1990; Nollert, 1992).

Protection

They are protected in a few countries such as Denmark, Germany, Hungary and Switzerland.

Description

The females of these green frogs grow to 9 cm in length (exceptionally 11 cm) and the males to 7 or 8 cm. The head is flat and triangular in outline with a short pointed snout. The hind legs are relatively short but the digits are very long with well-developed webs. They have smooth or slightly rough skin, green or bluish above with black blotches and a median stripe. The tip of the snout and upper jaw is almost always black. The limbs are well marked and the hind legs may have transverse bands. The whitish vocal sacs and yellow backs to the thighs differentiate these frogs from *Rana ridibunda*.

Distribution and habitat

These green frogs are found throughout the whole of Europe except the Iberian peninsula, and extend as far north as central Scandinavia. They occur as Far East as the River Volga. During the 18th, 19th and 20th centuries there have been attempts to introduce the species to Britain and some populations now thrive in parts of England.

It seems that they are euryoecious and therefore make use of a wide range of habitats including ponds of various sizes and conditions, gravel pits and even small lakes. They are essentially aquatic species and, even after spawning is complete, they remain in the water or in riparian habitats. It is common to find the species in the presence of other species of green frog.

Seasonal behaviour

They over-winter at the bottom of ponds. The young are active before the adults and emerge in early April or earlier in the southern parts of their distribution. Unlike other frogs, a period spent feeding precedes mating.

Vagility and population ecology

They spend more time in or near water than the Common Frog and do not range very far. They first breed when one or two years old but are not fully-grown until their fourth or fifth years.

Feeding ecology

These frogs tend to lie in wait for their prey rather than actively forage. The adults feed on a range of invertebrates but Hemiptera, Hymenoptera,

Dipterans and Coleoptera are mostly taken. Young frogs, newts and small fish are also sometimes eaten.

Thermal ecology

They are active during the day and exhibit behavioural thermoregulation. The weather conditions greatly influence calling behaviour, which usually ceases below about 12° C.

Reproduction, growth and development

Mating behaviour commences towards the end of May when males take up and defend small territories against rival males. The vocal sacs of the male lie in grooves behind the mouth but when extended appear as a pair of bladders the size of hazelnuts. Each time the male croaks the vocal sacs are inflated like small balloons. The vocalization is very loud and the chorus of a group of these frogs can be heard some distance away. These species mate at night and the spawn is deposited as several clumps of up to 1,000 eggs in each. The spawn is attached to vegetation in shallow water. The newly hatched larvae are brownish-black initially but soon acquire an olive colour with brown spots. Metamorphosis takes place after about 4 months. The fully developed larvae can be between 9 and 9 cm in length.

General

These species are not uncommonly found with other frog species, and much research has yet to be undertaken on resource partitioning and reproductive isolation amongst the different species.

Major references

Berger, L. (1977) Systematics and hybridisation in the *Rana esculenta* complex. In Taylor, D.H. & Guttman, S.I., eds. The Reproductive Biology of Amphibians. Pp. 367-388. Plenum Press, New York and London.

Forselius, S. (1963) Distribution and reproductive behaviour of *Rana esculenta* L. in the coastal area of N. Uppland, C. Sweden. Zoologiska bidrag fran Uppsala, 35, 517-528.

Gunther, R. (1990) Die Wasser Frosche Europas. Ziemsen, Wittenberg.

Nöllert, A. & C. (1992) Die Amphibien Europas. Franckh-Kosmos, Stuttgart.

Obert, H. –J. (1975) The dependence of calling activity in Rana esculenta Linn. 1758 and *Rana ridibunda* Pallas 1771 upon exogenous factors (Ranidae, Anura). Oecologia, 18, 317-328.

Tyler, M.J. (1958) On the diet and feeding habits of the edible frog (*Rana esculenta* Linn.). Proceedings of the Zoological Society of London, 131, 583-595.

Wahl, M. (1969) Untersuchungen zur Bio-Akustik des Wasserfrosches *Rana esculenta* (L.). Oecologia, 3, 14-55.

Wijnands, H.E.J. & Gelder, J.J. van (1976) Biometrical and serological evidence for the occurrence of three phenotypes of green frogs (*Rana esculenta* complex) in the Netherlands. Netherlands Journal of Zoology, 26, 414-424.

Rana lessonae / esculenta. **Adult.**

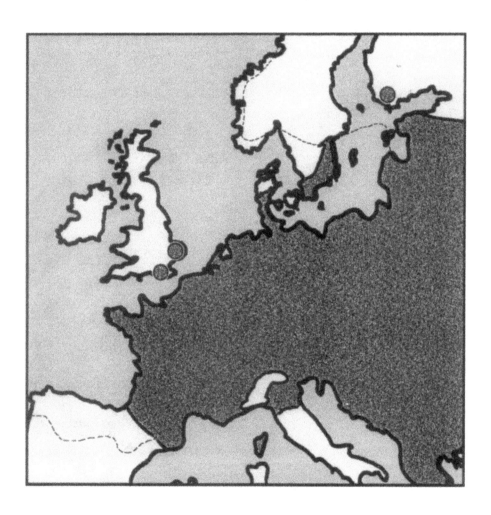

Rana lessonae / esculenta.

Family Ranidae

Rana ridibunda

Marsh, Laughing, Lake or Water Frog (GB), *Grenouille rieuse* (FR), Seefrosch (DE), Grote groene kikker (NE), Rana de Llanura (ES, Sjögroda (SE).

Introduction

This species and *Rana esculenta* are the 'green frogs' of Europe. They are very similar in size, colour and ecology but *R. ridibuna* is the larger of the two. Although their vocalizations are quite different, they do interbreed and the genetics of the resulting hybrids has been the subject of some very interesting studies. Green frogs may occur as 'pure' species and 'mixed' hybrid populations in many localities; in some places these two and other species may coexist.

Taxonomy

Ridibuna is Latin for 'laughing'. This species belongs to the group known as the *Rana esculenta* complex and is often found with *R. lessonae* and *R. esculenta*. It was previously considered to be a form of the Edible Frog. The nominate species has a broad geographical distribution and some congeneric species have now been recognised: *R. perezi* on the Iberian peninsula, *R. saharica* in Algeria and Egypt, and *R. epeirotica* and *R. shaeperica* from the Balkans. In some parts of northern Spain there are hybrids of *R. ridibunda* and *R. perezi*.

Protection

This species is protected in some countries such as Denmark, Germany, Hungary, Luxembourg and Switzerland.

Description

This is a large and robust frog – largest of all European frogs. It grows on average to about 15cm in length with females larger than males. The muscular vocal sacs are usually greyish and the general appearance of the skin can be rather warty. It is similar in size and colour to the Pool Frog and

Edible Frogs for which it is easily mistaken. However it has small metatarsal tubercles and longer legs; when the legs are folded alongside the body the heels overlap. Its colour varies from olive-green to brown above and green or greyish-white along the sides and on the thighs. It often has dark green or black spots.

Distribution and habitat

Although mainly an east European species, it has a broad geographical distribution across Europe from North Africa to western Asia. There are two main areas where it is found: in the southwest – Iberia and southern France – and in the east towards Russia. This species has been introduced to England where it occurs in the North Kent and Romney Marshes (thus the name Marsh Frog) and other localities in the south of England.

This semi-aquatic species is found in a variety of aquatic habitats in a range of environments including marshes, woodlands and farmland. For breeding it seems to prefer lakes, large ponds and even slow-flowing rivers. It appears to have been very successful in the brackish water of the Romney Marshes in southern England.

Seasonal behaviour

Temperatures of about 6° C will cause immobility and 4° C will induce winter torpor. It is therefore more sensitive to low temperature than other *Rana* species such as *R. temporaria* and *R. esculenta/lessonae*. In the north of its geographical range it over-winters from about October to March in water or in wetlands; in southern parts there may be no wintering period.

Vagility and population ecology

Generally it does not move far away from water but southern populations seem to be very agile and individuals may be found some distance from water. Population densities may be very high at 2,000 per hectare.

Feeding ecology

Adults forage both in water and on land and feed on a variety of invertebrates including spiders, flies and beetles. They also take much larger prey such as fish, salamanders, frogs, lizards and even small birds and mammals such as voles. The larvae feed mainly on algae, diatoms and worms.

Thermal ecology

R. ridibunda over-winters in the mud at the bottom of ponds until the water temperature rises to about 9° C. Egg laying takes place when the air temperature reaches between 15 and 25° C. It basks on the banks of ponds and lakes or on emergent vegetation, and tolerates high temperatures. With decreasing air temperature it changes its preference for water to land. Calling behaviour is influenced by temperature and below 12° C calling will cease. The onset, as well as the duration, of the reproductive period is temperature dependent.

Reproduction, growth and development

Marsh Frogs breed slightly earlier than closely related species. The reproductive period varies greatly in length; for example, in northern Greece it has been found to be very long and last for 135 to 165 days. In other localities a much shorter period for reproduction has been reported. It mates in April and is very noisy during the breeding season. The characteristic 'croaking chuckles' of the males can be heard throughout the summer. The weather seems to affect calling behaviour and they rarely call after days of rain or overcast weather. The breeding season may be prolonged with spawn being found late in the summer. The spawn is produced by day or night in several clumps 5-10cm below the water surface and always attached to vegetation; each clump contains several hundred eggs. Metamorphosis takes three months.

General

This species has a very variable reproductive period and thus seems well adapted to a wide range of conditions in the temperate climatic region of Europe. This aspect of its ecology, however, needs more data to allow a better understanding of the extent of that variation.

Major references

Borkin, L.J., Caune, I.A., Pikulik, M.M. & Sokolova, T.M. (1986) Distribution and structure of the Green Frog complex in the USSR. In Rocek, Z., ed., Studies in Herpetology, pp. 675-678. Charles University, Prague.

Huhn, B. & Schneider, H. (1984) Mating and territorial calls of the frog *Rana ridibunda* and their temperature-dependent variability. Zoologischer Anzeiger, 212, 273-305.

Kyriakopoulou-Sklavounou, xx & Kattoulas, M.E. (1990) Contribution to the reproductive biology of *Rana ridibunda* Pallas (Anura, Ranidae). Amphibia-Reptilia, 11, 23-30.

Menzies, J.I. (1962) The Marsh Frog (*Rana esculenta ridibunda Pallas*) in England. British Journal of Herpetology, 3, 43-54 & 89-90.

Sinsch, V. (1984) Thermal influences on the habitat preferences and the diurnal activity in three European *Rana* species. Oecologia, 64, 125-131.

Sinsch, V. (1991) Cold acclimation in frogs (*Rana*): microhabitat choice, osmoregulation, and hydromineral balance. Comparative Biochemistry and Physiology, 98A, 469-477.

Uzzel, T. (1975) Electrophoretic phenotypes of *Rana ridibunda*, *Rana lessonae*, and their hybridogenetic associate, *Rana esculenta*. Proceedings of the Academy of Natural Sciences, Philadelphia, 127, 13-33.

Rana ridibunda. **Adult.**

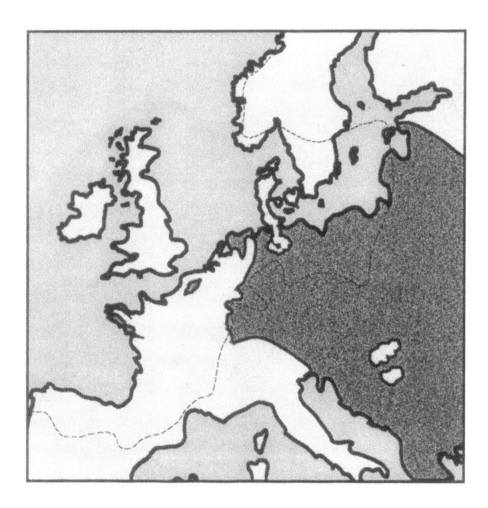

Rana ridibunda.

Family Ranidae

Rana temporaria

Common or Grass Frog (GB), *Grenouille rousse* (FR), Grasfrosch (DE), Bruine kikker (NE), Rana Bermeja (ES), Vanlig groda (SE).

Introduction

The Common Frog has been much exploited not least for laboratory use. The decline in its breeding sites and in numbers of individuals, possibly brought about by pesticides and habitat loss, has generated much concern. The Common Frog has had many admirers such as Percy Westell, who said in his delightful book, *British Reptiles, Amphibians and Fresh-water Fish:* "… lovers of wild nature should be able to appreciate this lower as well as the higher forms of creation. The bird which sings from the elm-tree ...... The Thrush inspires us with hope afresh and captivates the ear – the majestic sweep of the Buzzard – or the love-flight of the Skylark cannot fail to hold one entranced, but these creatures of the lowly earth, especially if they glide, creep, crawl or hop are looked upon with disdain. It's a poser for which I can't suggest an explanation".

Taxonomy

In addition to the nominate species, two sub-species have been described: *Rana temporaria honnorati* found in the foothills of the Alps and *R. t. parvipalmata* from northwestern Spain. The specific name originates in the word 'temporariss', which refers to the temporary occurrence of this species during breeding.

Protection

It is protected in a few countries such as Denmark, Germany, Hungary, Spain and Switzerland. In Britain it is illegal under the Wildlife and Countryside Act 1981 to offer Common Frogs for sale either dead or alive without a licence.

Description

The Common Frog has a blunt, short head, a fairly thickset body and a smooth, shiny skin covered with very small warts. It may grow up to 10 cm in length but is usually much shorter. Extremely variable in colour, it ranges from light yellow to reddish brown with dark markings to light orange with black markings. Albino forms are not uncommon and those with black nuptual pads have been recorded.

Distribution and habitat

This species is found throughout north and temperate Europe, Iberia and south to northern Italy. It is found up to 3,000 m and may occur near the snow line. In Britain it is widespread but not in the Outer Hebrides, Orkney and Shetland Isles. It was probably introduced to Ireland.

This species spends a lot of time out of water and is euryoecious, being found in a wide variety of wetland and moist habitats. Garden ponds probably make a significant contribution to the number of suitable breeding sites.

Seasonal behaviour

At low altitudes in the north of Spain it does not over-winter but in most localities it does, often near breeding ponds. The over-wintering period varies in duration depending largely on latitude. In northern Finland, for example, the Common Frog may over-winter for as long as 7 to 8 months compared to only four months in southern localities. This over-wintering is not a state of hibernation because the frogs remain reactive and may even move about.

The annual migration has been much researched but there is some uncertainty about the cues used by frogs: chemical odours, celestial bodies and the earth's magnetic field have been suggested.

The Common Frog is capable of olfactory orientation: it can detect chemical odours, but in one study of its navigational capability it seemed the smell of a pond and the earth's magnetic field were not primary orientational cues.

Vagility and population ecology – The movement of frogs about a pond seems to be quite random and, although some individuals show some site preferences, there does not seem to be any evidence to show that they are territorial.

Feeding ecology

The Common Frog feeds mainly on land and takes a whole range of invertebrates, especially molluscs, woodlice, beetles and insect larvae. Some studies have shown that as much as 15% of food taken is made up of plant material. This could have been eaten by accident but it is perhaps more likely that plant material is indeed part of its normal diet. The tadpoles graze on algae and take small crustaceans such as daphnia.

Thermal ecology

> Many will over-winter in the mud at the bottom of ponds where water temperatures may drop to 0° C. During this wintering period, and usually during spawning, they eat nothing and exist on nutritional reserves, mainly glycogen, built up over the previous summer. During the summer Common Frogs prefer temperatures around 10 to 20° C. The spawn will not survive high temperatures, and in one study it was found that the eggs soon perished at temperatures above 24 to 25° C.

Reproduction, growth and development

> The Common Frog is an 'explosive' breeder: breeding is generally synchronous and for any one site takes place only over about ten days. Males arrive at the pond before females, and the larger males tend to aggregate at spawning sites, while the younger ones search for females around the edge of ponds. Male mating success, however, has been found to be random with regard to body size, and in some localities small males are more successful. The call of the Common Frog and chorusing are well known but it seems there is more than one kind of call. One study distinguished three calls on the basis of call duration and rate. Two of these calls may help to prevent contact between males and the third main optimum dispersion when chorusing.
>
> Variation in the timing of breeding depends on altitude and latitude: in northwest Spain at low altitudes reproduction takes place between November and February. In other localities Common Frogs spawn in late January though others may not do so till early June. Egg numbers per egg mass range on average from several hundred to a few thousand but typically are around 1,000. There is much variation in egg numbers produced due mostly to the size of the female frog and environmental conditions. The average time between deposition of spawn and hatching is between 18 and 23 days and the percentage of eggs that hatch ranges from 92 to as low as 20%. There is considerable weight loss in both sexes – in the range of 0.5% of total body weight per day – during and after the breeding season until about May.
>
> The newly emerged tadpoles have well-developed gills and for the first few days move about very little. Metamorphosis can be complete after about 10 to 12 weeks but often takes longer.

General

> The daily aggregation of tadpoles along the edges of ponds would at first seem to be some kind of thermoregulation. In the Common Frog the cycle of aggregation and movement to deeper parts of the pond are synchronized

to daily cycles of light intensity and temperature but aggregations may occur even when there is no temperature gradient at the edge of the pond. Aggregations could be caused by an abundance of algae along pond edges but it is more likely that this behaviour is caused by several factors.

Major references

Gelder, J.J.van, Evers, P.M.G. & Maagnus, G.J.M. (1978) Calling and associated behaviour of the Common Frog, *Rana temporaria*, during breeding activity. Journal of Animal Ecology, 47, 667-676.

Itamies, J. & Koskela, P. (1970) On the diet of the common frog (*Rana temporaria* L.). Aquilo Serie Zoologica, 10, 53-60.

Koskela, P. & Pasanen, S. (1974) The wintering of the common frog, *Rana temporaria* L., in northern Finland. Aquilo Ser. Zool., 15, 1-17.

Mensi, P., Lattes, A., Macario, B., Salvidio, S., Giacoma, C. & Balletto, E. (1992) Taxonomy and evolution of European brown frogs. Zoological Journal of the Linnean Society, 104, 293-311.

Miaud, C., Guyetant, R. & Elmberg, J. (1999). Variations in life-history traits in the common frog *Rana temporaria* (Amphibia: Anura): a literature review and new data from the French Alps. Journal of Zoology (London), 249, 61-73.

Muller, W. & Kiepenheur, J. (1976) Olfactory orientation in frogs. Die Naturwissenschaften, 63, 49.

Reading, C.J. (1984) Interspecific spawning between common frogs (*Rana temporaria*) and Common toads (*Bufo bufo*). Journal of Zoology, 203, 95-101.

Ryser, J. (1989) The breeding migration and mating system of a Swiss population of the common frog *Rana temporaria*. Amphiba-Reptilia, 10, 13-21.

Sinsch, V. (1984) Thermal influences on the habitat preference and diurnal activity in three European *Rana* species. Oecologia, 64, 125-131.

Rana temporaria. **Adult.**

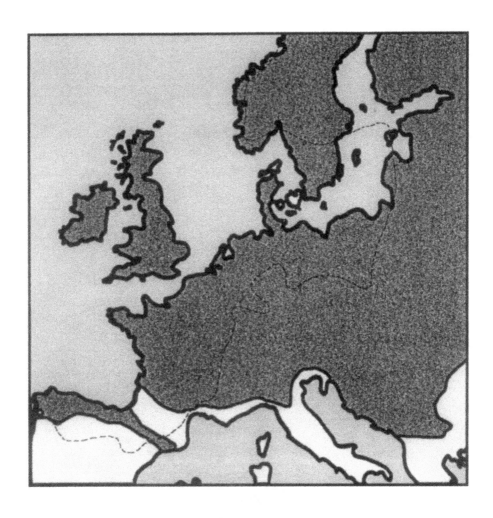

Rana temporaria

Family Anguidae

Anguis fragilis

Slowworm or Blindworm, Legless Lizard (GB), Orvet (FR), Blindschleiche (DE), Hazelworm (NE), Lución (ES), Kopparödla, Ormslå or Kopparorm (SE).

Introduction

The Slowworm is a favourite of many young naturalists; a lizard without legs and an example of the subtle difference between snakes and lizards. Despite its popularity with children, much of the ecology and behaviour of the Slowworm is still a mystery. Early references made claim to a diet of snails and worms and it was once said that "though its teeth are too insignificant to penetrate, the Slow worm is very savage and bites furiously".

Taxonomy

The generic name comes from the Latin *Anguis* meaning serpent and *fragilis* meaning brittle or fragile, a reference to its ability to cast all or part of its tail (tail autotomy) when handled or treated in a rough manner. In addition to the nominate species, two sub-species have been described: *Anguis fragilis colchicus* from the eastern areas of its geographical distribution, and *A. f. peloponnesiacus*, which is endemic to the Peloponnese peninsula, Greece. Some populations to the Peloponnese have been described as a new species, *Anguis cephalonicus*.

Protection

This lizard is often found within urban areas and on disturbed land where it can be abundant in a few localities. Anecdotal observations suggest that there have been sharp declines in population densities, thus giving some cause for concern and a case for its better protection. It is widely protected throughout Europe, and in Britain it is illegal to harm them and offer Slowworms for sale under the Wildlife and Countryside Act 1981.

Description

Many specimens have broken or re-grown tails – the lizard as a whole does not break in two as quaintly noted in Bell's *British Reptiles of 1839*. Smith (1990) recorded damaged tails in 50% of adults and 20% of juveniles. The undamaged tail is slightly longer than the body. In Europe adults have been known to reach a total length of over 52 cm but usually they are much smaller. The snout-vent lengths of adults from populations in mainland England have been found to be up to 20 cm, and in populations of off-shore islands to have an s-v length as great as 25 cm.

Males are uniform in colour, greyish brown with very smooth scales; in some localities they may have blue spots. Females are usually black underneath, light brown above with a dark vertebral stripe and dark sides. The colour and markings of female Slowworms has been compared to that of Adders (*Vipera berus*) with the unlikely suggestion that they are able to mimic adders and thus gain protection. Newborn Slowworms are dark brown or black beneath, a striking pale golden brown above with a dark vertebral stripe. The blue spotting of Slowworms makes them most attractive but there is only one detailed study of what has yet to be established as a true polymorphism.

Distribution and habitat

Anguis fragilis is found throughout Europe except Ireland, southern Scandinavia and parts of southwest Asia. It is also found on many islands off the coast of Europe. This lizard occurs in a wide range of habitats – from lowlands to altitudes of 2,300 m in Austria – but mostly commonly among low lying temperate grasslands on base-rich soils. It does not seem to tolerate very short grass, intensive agriculture, dense forest and woodlands.

Seasonal movement and behaviour

Emergence from over-wintering occurs during March and the males are the first to appear and not uncommonly bask in the early spring sunshine among grassy tussocks. Females and juveniles emerge some weeks later during April but rarely bask in the open. There is very little information about the time of mating but it seems to occur any time between April and June. Newly born young can be found from the end of August onwards, and by the end of October both adults and young have sought the protection of winter refuges

Vagility and population ecology

The secretive· behaviour of Slowworms and the difficulty of identifying individuals has, until recently, prevented studies of their movements. However, making use of photographic records of ventral neck markings in a study lasting many years in England has provided much detailed information about their movements. The first long-term study has revealed that Slowworms have a mean home range area of about 270 square metres for males and 170 square metres for females. Distances between day-to-day recapture points for individual lizards have been found to be about two metres.

Slowworms are long-lived animals and there are records of some as old as 20 years or more. Smith (1951) quotes one from Copenhagen that lived for at least 54 years.

Feeding ecology

Studies in the field and in captivity have revealed that slugs, worms and other small invertebrates, such as spiders, beetles, millipedes and woodlice, are common items of prey. Reports from Italy suggest that Slowworms may take pseudoscorpions as well as spiders and beetles. In Spain the prey includes fly larvae as well as woodlice and millipedes.

Thermal ecology

The mean body temperature for normal activity of Slowworms in captivity has been found to be 23° C (14-29° C). In one study of 23 adults in England and Wales the mean body temperature of animals in the field during optimal conditions was found to be 22.6° C (14.5-28.0° C). In a much more extensive study carried out over some years (174 measurements) the overall mean body temperature was found to be 26.6° C and interestingly females had higher body temperatures than males. Although male Slowworms do occasionally bask during the spring and females later in the year, they spend most of the time hidden below the litter of vegetation. They are commonly found beneath debris – sheets of tin or pieces of board lying on the ground – and it is there they are able to 'bask' in the heat yet not in direct sunlight.

Reproduction, growth and development

The males are sexually active from May to June and evidence of mating – mating scars on females – has been found from mid-May onwards. The young are born from August to September and seasonal weather conditions may have a marked effect on gestation time. Typically eight young are born and the average weight is 0.5 g. In Britain evidence has been found to suggest that female Slowworms do not breed in consecutive years and that biennial breeding is the norm. They do not seem to feed in the latter stages of pregnancy and may be emaciated after giving birth; the following summer is therefore used to put on weight. In dry summers, however, slugs and earthworms may be difficult to find and for this reason dry summers are thought to affect reproductive success.

In captivity, newly born lizards have a growth rate (based on lengths) similar to the lacertids *Lacerta vivipara* and *Podarcis muralis*. Later growth rates may be quite variable. Young animals of about 11 cm increase in length by about 1.5 to 2.0 cm in a year. The males seem to become mature when about 14 cm long and about 4 years old; females at about 15 cm in length and 5 years old.

General

This species is found on the mainland and also on a number of offshore islands around Britain where they may have been isolated for 7,000 to 8,000 years. No detailed studies have been undertaken on the island populations of *Anguis fragilis* and therefore an interesting area of investigation would be a comparative study of the ecology and genetics of island populations.

Major references

Dzukic, G. (1987) Taxonomic and biogeographic characteristics of the slow worm (*Anguis fragilis* Linnaeus 1758) in Yugoslavia and on the Balkan Peninsula. Scopolia, 12, 1-46.

Halton, C. & Avery, R.A. (1989) Growth dynamics of juvenile slowworms (*Anguis fragilis*) in captivity. British Herpetological Society Bulletin, 30, 19-21.

Patterson, J.W. (1983) Frequency of reproduction, clutch size and clutch energy in the lizard *Anguis fragilis*. Amphiba-Reptilia, 4, 195-203.

Patterson, J.W. (1990) Field body temperatures of the lizard *Anguis fragilis*. Amphibia-Reptilia, 11, 295-306.

Platenberg, R.J. & Griffiths, R.A. (1999) Translocation of slow-worms (*Anguis fragilis*) as a mitigation strategy: a case study from South-East England. Biological Conservation 90, 125-132

Poivre, C. (1975) Observations sur le comportment predateur de l'orvet (*Anguis fragilis* L.). II. Capture de diverses proies. Terre vie, 29, 63-70.

Smith, N.D. (1990) The Ecology of the Slowworm (Anguis fragilis L.) in Southern England. 157 pp. M.Phil. thesis, University of Southampton.

Stumpel, A.H.P. (1985) Biometrical and ecological data from a Netherlands population of *Anguis fragilis* (Reptilia, Sauria, Aguidae). Amphibia-Reptilia, 6, 181-194.

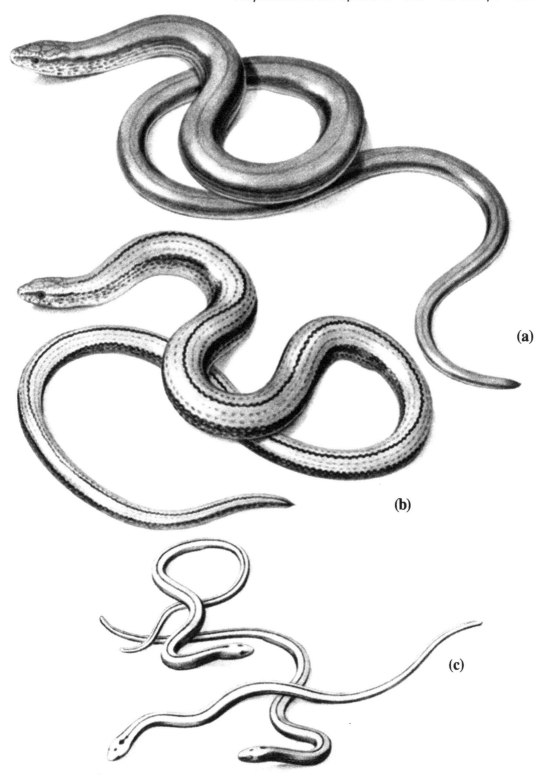

Anguis fragilis. **Male (a), female (b), juveniles (c).**

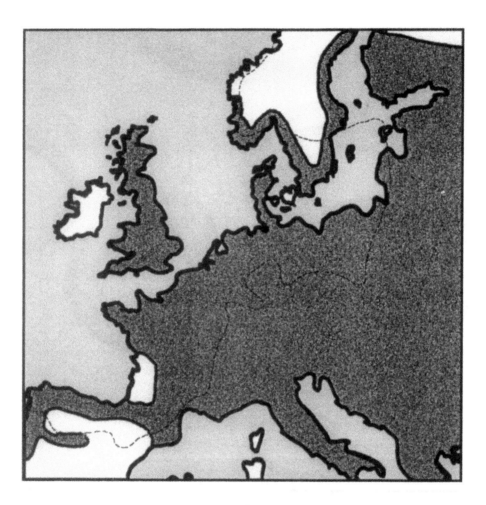

Anguis fragilis.

Family Lacertidae

Lacerta agilis

Sand Lizard (GB), Lézard des souches (FR), Zauneidechse (DE), Zandhagedis (NE), Lagarto Agil (ES), Sandödla (SE).

Introduction

This lizard is well known in temperate Europe and has attracted the attention of naturalists for decades. In 1862, Oliver Goldsmith described the Sand Lizard: "This species is found on sandy heaths in some parts of England. It is more sluggish than the common lizard, and will attempt to bite anyone who handles it."

Taxonomy

The specific name means agile or busy – a feature of many lacertid lizards. There have been many sub-specie described and an important taxonomic feature is the head scalation.

Protection

The reduction in the distribution and decline in population densities of this lizard has been well documented. It is listed in Appendix II of the Berne Convention and is protected in most countries; in Britain, for example, it is protected under the Wildlife and Countryside Act 1981 (Variation of Schedules 1988, 1991).

Description

Adult Sand Lizards are larger and more robust than Viviparous Lizards. Typical adults are about 6 to 8 cm in s-v length and weigh about 10 to 12 g. The tail is about 1.5 times the body length. Males have slightly larger heads than females and during the spring are distinguished from females by intense green-coloured flanks. The colour of both sexes is variable and typically the Sand Lizard is greyish or brown but has noticeable ocelli (white spots with dark rings) and brown or reddish bands. Newly born Sand Lizards have striking ocelli covering most of their bodies.

Distribution and habitat

Sand Lizards have a fragmented distribution throughout the temperate climatic region, inhabiting open deciduous woodland, heathlands, grasslands, sand dunes, hedges and roadside verges. Two important features of the habitat seem to be dense ground cover and conditions where the females can excavate burrow for their eggs. A detailed study of Sand Lizard habitats in southern England revealed that populations were associated either with heath communities, in various stages of succession to woodland, or sand-dune communities. The highest population densities were found in the most structurally diverse habitats with mixtures of different vegetation heights, with banks and gullies, and generally in the most sunny locations.

Seasonal movements and behaviour

Emergence from over-wintering occurs towards the end of March with the males typically appearing first. Mating takes place in May, both sexes showing a variety of postures and movements before actual mating takes place. On occasions, the males may chase and fight each other but the Sand Lizard is not a territorial species. In June or July, females devote considerable time to the selection of nest sites, construction of nests in shallow burrows with crescent-shaped entrances, and laying the eggs. Newly hatched Sand Lizards are typically seen during August and then by mid-October both adults and young have retreated to their wintering sites. Nothing is known about their winter behaviour or wintering sites.

Vagility and population ecology

Typically, this species is found in discrete populations and although individuals may on occasions migrate over some hundreds of metres, they tend to keep within fairly restricted areas. Until more recently, studies of their movements were very difficult because of the problems of continuous observation of individuals in sometimes very dense undergrowth. The use of very small radio transmitters attached to large adult lizards gave new insights to their way of life. The home range of a Sand Lizard varies amongst individuals and on the type of habitat. Typically the area is several times greater than that of a cricket wicket and may sometimes be nearly 2,000 square metres.

In Britain, assessments of countrywide population size have been attempted but such assessments have little value because so many assumptions had to be made. Estimates of local population size, however, can be based on sightings of individual lizards, using the characteristic markings on the individual's neck and back. It may take at least two years of intensive field studies before the population size can be estimated with any degree of confidence.

It is more difficult to determine changes in population size, rates of recruitment and changes in density because of the difficulties of knowing whether most individuals in the population have been sighted. The tendency to live in fairly discrete populations does however make it possible to monitor changes in the number of populations and their geographical distribution. Unfortunately, the number of populations, which have become extinct during the last two decades, is far greater than the number of newly discovered populations. The isolated nature of many of the remaining populations gives great cause for concern as regards the survival of this species.

Feeding ecology

Sand Lizards feed on a variety of invertebrate prey of different sizes, most common taxa being beetles, spiders, flies and ants. They go in search of prey and have been known to stalk and capture large butterflies and beetles. There are reports of adults killing and eating juvenile lizards but it is not known if this affects population size or what causes such behaviour – possibly crowding or lack of food could prompt cannibalism. Although the feeding ecology of this species has been well researched there remain some unanswered questions: it is not known, for example, which kind of habitat or which conditions provide the best food sources for the Sand Lizard throughout the year. Heathlands can support high species richness of spiders and insects and it is likely that mixtures of grasslands, scrub and heathland would support high levels of prey species for long periods throughout the year.

Thermal ecology

The average body temperature during normal activity is 31° C and the critical minimum temperature for populations in England is about 5° C in summer and about 3° C in winter. Although the activity temperature for Sand Lizards in Germany is the same, the critical minimum temperature is lower. The Sand Lizard spends comparatively more time basking and has greater heat requirements than the Viviparous Lizards: respectively temperatures of about 26° C on basking surfaces such as logs as opposed to 23° C. It is not surprising that they tend to be found in south-facing sheltered areas.

Reproduction, growth and development

The nest sites are selected by the females. South-facing slopes inclined at 30° or less seem important for successful incubation. The average clutch size is 5 to 6 eggs – old adults may lay as many as 13 eggs – and these are surprisingly large, about 15 mm long. Eggs laid at a depth of 7 to 10 cm have the greatest chance of surviving. After the eggs are laid the female covers and abandons them. The incubation period of between 55 and 70 days is temperature dependent and during this time the eggs are at risk from predators and disturbance.

General

The biology of the Sand Lizard has been well researched and this knowledge has had important applications in the conservation and management of the species. The ecology of habitat restoration for this species needs further work.

Major references

Corbett, K.F. & Tamarind, D.L. (1979) Conservation of the sand lizard *Lacerta agilis* by habitat management. British Journal of Herpetology, 5, 799-823.

Dent, S. (1986) The ecology of the Sand Lizard *Lacerta agilis* L. in forestry plantations and comparisons with the Common Lizard *Lacerta vivipara* Jacquin. Ph.D. thesis, University of Southampton.

Glandt, D. & Bischoff, W. (1988) Biologie und Schutz der Zauneidechse (*Lacerta agilis*). Mertensiella (Supplement zu Salamandra), 1, pp. 257.

House, S.M. & Spellerberg I.F. (1983) Ecology and conservation of the Sand lizard (*Lacerta agilis*) habitat in southern England. Journal of Applied Ecology, 20, 417-437.

Jabolokov, A. (1976) The Sand Lizard. *Lacerta agilis*. 376 pp. Nauka, Moscow. (in Russian)

Korsos, Z. & Gyovai, F. (1988) Habitat dimension and activity pattern differences in allopatric populations of *Lacerta agilis*. Mertensiella. 1, 235-244.

Nicholson, A.M. (1980) The ecology of the Sand Lizard (*Lacerta agilis* L.) in southern England and comparisons with the Common Lizard (*Lacerta vivipara* Jacquin). Ph.D. thesis, University of Southampton.

Nollert, a. (1989) Beitrage zur Kenntnis der Biologie der Zauneidechse, *Lacerta agilis argus* (Laur.), dargestellt am Beispiel einer Populationer aus dem Bezirk Neubrandenburg. Zoologische Abhandelungen Staatliches Museum fur Tierkunde Dresden, 44, 101-132.

Nuland, G.J.van & Strijbosch, H. (1981) Annual Rhythmics of *Lacerta vivipara* Jacquin and *Lacerta agilis agilis* L. (Sauria, Lacertidae) in the Netherlands. Amphibia-Reptilia, 2, 83-95.

Strijbosch, H. & Creemers, C.M. (1988) Comparative demography in sympatric populations of *Lacerta vivipara* and *Lacerta agilis*. Oecologia, 76, 20-26.

Lacerta agilis. **Male (a), female (b).**

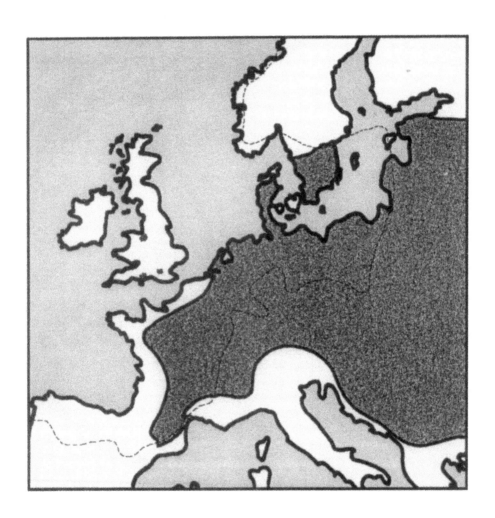

Lacerta agilis.

Family Lacertidae

Lacerta viridis

Green or Emerald Lizard (GB), Lézard vert (FR), Smaragdeidechse (DE), Smaragdhagedis (NE), Lagarto Verde (ES), Smaragdödla (SE).

Introduction

The lizards *Lacerta viridis* and the smaller *Lacerta agilis* are two well-known examples of 'green lizards'. They occur together (sympatric) in some parts of their distribution, and the habitats of the two species are similar. Other species, which make up this group of medium to large green lizards in Europe, are *Lacerta schreiberi* and *Lacerta trilineata.*

Taxonomy

The species name viridis is Latin for green. A number of sub-species are now recognized.

Protection

Although it is considered to be a persecuted species in many countries, its habitat is not protected in all of Europe. There are some relict populations, in Germany for example, and it is those that urgently require protection. *Lacerta viridis* is protected in France, Germany and Hungary, and is listed in Appendix II of the Berne Convention.

Description

This lizard has a thickset body and is the largest of all lizards found to the north of the alpine region. It grows to about 13 to 14 cm in body length. The tail is noticeably long, very often at least twice the body length, and so the total length may be between 30 and 40 cm. The males are usually entirely green with obvious fine black stippling and a darker light-spotted head. The underside of the head and throat is often coloured vivid blue especially during the mating period. The females are more variable in colour, sometimes entirely green but some with shades of brown. Adult females sometimes have a blue tinge on the throat, two or four white stripes along their flanks and black spots on the upper side. The underneath of both sexes is usually yellowish-green. The young are usually brownish or beige with a few light spots.

Distribution and habitat

This species is found throughout Europe, including the Channel Islands, and as far south to Spain, Italy and Greece; few Mediterranean islands are known to have it. There have been unsuccessful attempts to introduce it to various parts of Britain.

There has been surprisingly little ecological research on green lizard habitat but there are some general reports. *L. viridis* is found mostly in sunny, sparsely wooded areas, shrub-dominated landscapes, or grassland with some brambles, gorse and bracken; also dense hedgerows and overgrown embankments. There is a tendency for it to occur in slightly denser and more structurally complex woodlands and shrublands than *L. agilis*. In some regions, however, such as along the River Rhine in Germany, there is a danger that some habitats are becoming overgrown with too much vegetation.

Where *L. viridis* and *L. agilis* are sympatric, it seems likely that the two species utilize soil, air temperatures and exposure to light in different ways. Yet, apart from differences in periods of activity, there are apparently only very subtle differences between the two species.

Seasonal movements and behaviour

Emergence from the over-wintering sites occurs in March to April but is temperature dependent. German studies, for example, in Brandenburg (east) and the Rhineland-Palatinate (west) have shown that western populations of *L. viridis* commence daily activity at temperatures at least 7° C lower when compared to eastern ones, and as a result the annual activity begins two or three weeks earlier.

Studies of sympatric populations of *L. viridis* and *L. agilis* have found differences in their daily activity rhythms. The larger Green Lizard has two foraging peaks and extends its activity further into the late afternoon, especially during the summer. The Sand Lizard was found to have a unimodal activity and appeared earlier in the morning. Temperature is the main factor determining this activity but which may be modified by sexual and physiological condition.

Vagility and population ecology

Young animals tend to form groups before over-wintering and isolated lizards eat less and grow less well than other lizards in-groups. In general and in suitable habitats, an individual's home range is about 30 to 50 m in diameter. The numbers of individuals in each age-class can vary immensely: after unfavourable summers as many as 85% of lizards may be more than 2 years old but occasionally with good recruitment that can fall to 50%.

Feeding ecology

L. viridis forages for its prey amongst dense herbaceous vegetation and under the edges of shrubs. It frequently forages from shrubs into nearby grassland and fairly open areas. On some occasions it has been known to hunt while climbing on bushes and small trees – when high in the stems of plants the long tail is used to spread the animal's weight. Its diet is typically broad and includes many kinds of invertebrates in both larval and adult form. It is also known to feed on small fruits and the eggs of some small birds.

Thermal ecology

The Green Lizard has a range of basking postures to aid efficient heat uptake. Studies have shown that there is a distinct annual cycle in its day and night voluntary body temperatures. From about May through to July the lizards have high body temperatures of 33° C and there is little diel variation. As the period of over-wintering approaches, the diel cycle of body temperatures increases, which is most likely to be governed by changing photoperiods.

Reproduction, growth and development

Breeding age is reached after the second winter. Mating occurs about three weeks after emergence from over-wintering. The eggs are laid about four to six weeks after mating, in May or June. In some localities, such as northwest France, there are two phases of mating and subsequently of egg-laying, one in May and one in June. Between 6 and 21 eggs are laid and the incubation time, being temperature dependent, takes between 2.5 and 3.5 months.

General

In any one year not all adults will breed and social factors may partly determine which individuals do so.

Major references

Böker, T. (1990) Zur Okologie der Smaragdeidechse *Lacerta viridis* (Laurenti, 1768) am Mittelrhein. I. Lebensraum. Salamandra, 25, 19-44.

Böker, T. (1990) Zur Okologie der Smaragdeidechse *Lacerta viridis* (Laurenti, 1768) am Mittelrhein. II Populationsstruktur, Phenologie. Salamandra, 26, 97-115.

Korsos, Z. (1982) Field observations on two lizard populations (*Lacerta viridis* Laur. and *Lacerta agilis* L.). Vertebratea Hungarica, 21, 185-194.

Korsos, Z. (1984) Comparative niche analysis of two sympatric lizard species (*Lacerta viridis* and *Lacerta agilis*). Vertebrata Hungarica, 22, 5-14.

Perkins, C.M. (1988) The Green Lizard and Wall Lizard in Jersey: distribution, biology and conservation. M.Sc. thesis, University of Bristol.

Peters, G. (1970) Studien zur Taxonomie, Verbreitung und Ocologie der Smarag-deidechsen. IV. Zur Okologie und Geschichte der Populationen von *L. v. viridis* (Laurenti) in mitteleuropaischen Flachland. Beitrage zur Tierwelt der Mark VII. Veroff. Bezirksheimatmuseums Potsdam, 21, 49-119

Rismiller, P.D. (1987) Thermal Biology of *Lacerta viridis*: seasonal aspects. Doctoral thesis, Philipps-Universitat Marburg/Lahn.

Saint Girons, H., Castanet, J. & Bradshaw, J.P. (1989) Demographie comparee de deux populations francaises de *Lacerta viridis* (Laurenti, 1768). Revue Ecologie (Terre Vie), 44, 361-386.

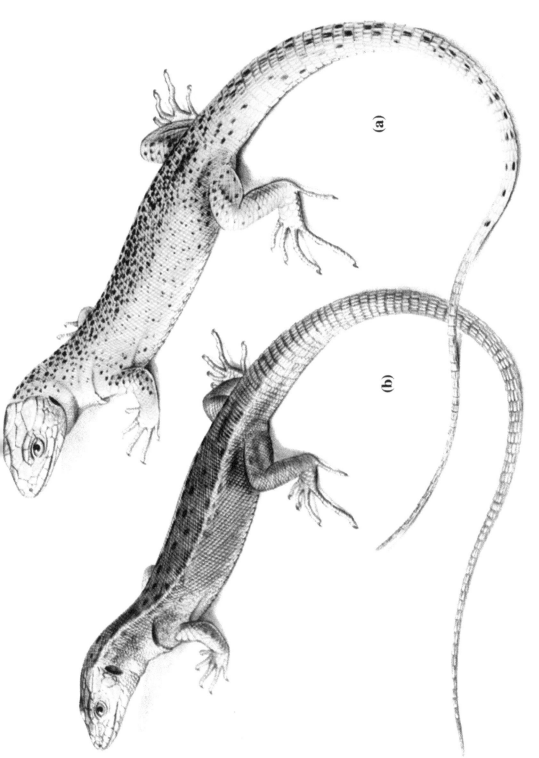

Lacerta viridis. Male (a), female (b).

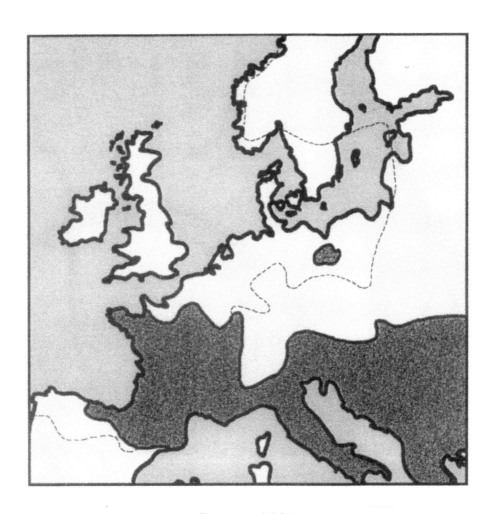

Lacerta viridis.

Family Lacertidae

Lacerta vivipara

Common or Viviparous Lizard (GB), Lézard vivipare (FR), Waldeidechse oder Bergeidechse (DE), Levendbarende hagedis (NE), Lagartija de Turbera (ES), Skogsödla (SE).

Introduction

The Common Lizard has been described as a "pretty, active little creature" and "a frequent inhabitant of our heaths". The natural historian Baron Joseph von Jacquin, 1766-1839, a friend of Mozart, first described the Common Lizard as long ago as 1787. Not much seems to be known about Jacquin but it seems he had an interest in birds and that he described the type specimen of *Lacerta vivipara* from Vienna when he was about 21.

Taxonomy

The specific name *vivipara* means live-bearing, the mode of reproduction which occurs in most of the populations. Until recently no sub-species had been described, though *Lacerta vivipara pannonica* from the lowlands of eastern Slovakia has now been recognized.

Protection

Although this species remains fairly abundant, there are many localities where it was once widespread but has now vanished. It is protected in many countries and in Britain, for example, it is illegal to sell or harm Common Lizards under the Wildlife and Countryside Act 1981. They are fully protected in Northern Ireland.

Description

Delicate and slender in appearance, the Common Lizard is brown and grey or dark olive-green with brown and black markings. Adult males are easily distinguished from adult females by being on average slightly smaller but with relatively longer tails and legs. During early summer breeding males are dark orange on their ventral surface, and they have swellings at the base of the tail caused by hemipenes. Larger specimens have an s-v length of about 6 cm and tails almost twice as long as the body. Variation in morphology has been found to be greater in lowland areas than in more northern or more mountainous regions. The new-born lizards are very dark coloured and occasionally melanic adults are found. A number of specimens with deformed or double tails have been found, and it is not unusual to find animals with re-grown tails.

Distribution and habitat

The Common Lizard has an ubiquitous distribution: habitats range from woodlands, heathlands and sand dunes through roadside verges and hedges to urban gardens. On inland dunes in the Netherlands, habitats with a strong spatial complexity seem to be preferred, no matter how humid or dry. In the south, the Common Lizard occurs up to about 3,000 m where it is found associated with wet meadows, marshes or streams.

Even more remarkable than the wide variety of habitats is the fact that there is only one sub-species, yet *L. vivipara* has a very wide geographical distribution throughout the northern and central Palearctic, including Europe, Scandinavia to the Arctic, through northern Asia to the Pacific coast.

Seasonal movements and behaviour

One of the first reptiles to emerge each spring, male Common Lizards may be seen as early as mid-February. Mating occurs during April and young are born any time from July to August. By the end of October, most Common Lizards will have sought shelter but some juveniles may remain for some further weeks. They over-winter in tree trunks, plant litter and beneath logs and stones.

Vagility and population ecology

Common lizards do not seem to live in well-defined colonies, nor do they seem to show any great site tenacity. In one homing behaviour experiment where lizards were displaced at varying distances from the place of capture, only 50% returned when displaced 70 m and as few as 29% by 100 m. Mortality in the first year of life can be as high as 90%. The Common Lizard's longevity is only four to five years and most adults are likely to die before their fourth winter although some my survive until 8 years old. The life expectation of males is higher than for females. Snakes are among the many predators. Common Lizards are able to detect pheromones left by snakes such as Adders and Smooth Snakes.

Changes in density affect a population's structure, particularly the mortality of juveniles and the number of young per female (affected also by body size).

Feeding ecology

In a study of this species in Britain, a wide variety of insects and other invertebrates was taken but spiders were found to be the principal food with Homoptera (bugs such as leaf hoppers) being important during the summer. There seems to be little difference between the prey taken by both adults and young although smaller items tended to be taken by the younger lizards. Spiders and Homoptera have also been found to be the principal prey for *L. vivipara* in other countries.

Thermal ecology

The voluntary body temperature, as measured in the laboratory, has been found to vary throughout its period of activity – 27.3 to 32.4° C. In general it is higher in the spring and autumn than in the summer although for females it was found to be very low in early spring. Thermoregulation in the field can be very precise and the lizards tend to select mostly dry grass for basking sites when it is warm but logs when there are periods of cloud cover interspersed with sunny periods. Their sprinting speed is highly temperature dependent and rises rapidly between 20 and 32° C. Tongue-flicking rates are also temperature dependent but the chemical deposits left by predators, such as Adders, will affect both tongue-flicking and basking behaviour.

Detailed physiological studies of *L. vivipara* have shown that it seems able to adjust and conserve its levels of energy consumption when body temperature is within the voluntary range. During the winter it is able to diminish levels of energy expenditure below expected levels and seems to have specialised physiological adjustments for over-wintering in addition to the typical ability to store fat.

Reproduction, growth and development

Sexual dimorphism by colour in *L. vivipara* is fairly well pronounced and it has been found that colour and also odours are important cues during courtship. In general mating takes place during April and May and gestation takes about three months. Between five and eight young are born (average 7.7) but larger litters have been recorded. The length of the new-born is about 2.0 cm (s-v). In most localities, this lizard gives birth to live young: more correctly, the young are born enveloped in a membranous egg which is soon ruptured by the emerging young. However at Bagneres-de-Bigorre in the Central Pyrenees, *L. vivipara* has been known to lay eggs, and in the Cantabrica Mountains of Spain populations of this lizard are permanently oviparous (egg laying). In the north of Spain there is prolonged retention of the eggs until a fairly advance stage of development has been reached –

eggs are laid in late July. Sexual maturity is reached before the second winter and under favourable conditions in the southern parts of its range, as many as 50% may be able to breed when they are only a year old.

General

Clearly the little Common Lizard is a very successful animal and has adapted to a wide variety of climatic conditions and habitats. It is not a specialist predator, it uses microclimate with great success, it has specialized adaptations to survive the winter and it can be either viviparous or oviparous. Furthermore, it seems able to distinguish between the odours left by those snakes which feed on it, *Vipera berus* and sometimes *Coronella austriaca*, and *Natrix natrix* which does not.

Major references

Avery, R.A. (1966) Food and feeding habits of the Common Lizard (*Lacerta vivipara*) in the west of England. Journal of Zoology, 149, 115-121.

Damme, R.Van, Bauwens, D. & Verheyen, R.F. (1987) Thermoregulatory responses to environmental seasonality by the lizard *Lacerta vivipara*. Herpetologica, 43, 405-415.

Heulin, B. & Guillaume, C. (1989) Extension geographique des populations ovipares de *Lacerta vivipara*. Revue Ecologie (Terre et Vie), 44, 39-45.

Heulin, B., Arrayago, M.J., Bea, A. & Brana, F. (1992) Caracteristiques de la coquille des oeufs chez la souche hybride (ovipare x vivipare) du lézard *Lacerta vivipara*. Canadian Journal of Zoology, 70, 2242-2246.

Massot, M., Clobert, J., Pilorge, T., Lecompte, J. & Barbault, R. (1992) Density dependence in the Common Lizard: demographic consequences of a density manipulation. Ecology, 73, 1742-1756.

Patterson, J.W. & Davies, P.M.C. (1978) Energy expenditure and metabolic adaptation during winter dormancy in the lizard *Lacerta vivipara* Jacquin. Journal of Thermal Biology, 3, 183-186.

Pilorge, T. (1981) Determination ode l'age dans une population naturelle du Lézard vivipare (*Lacerta vivipara* Jacquin 1787). Acta Oecologica, 2, 3-16.

Strijbosch, H. (1988) Habitat selection of *Lacerta vivipara* in a lowland environment. Herpetological Journal, 1, 207-210.

Strijbosch, H. & Creemers, R.C.M. (1988) Comparative demography of sympatric populations of *Lacerta vivipara* and *Lacerta agilis*. Oecologia, 76, 20-26.

Lacerta vivipara. **Male (a), female (b).**

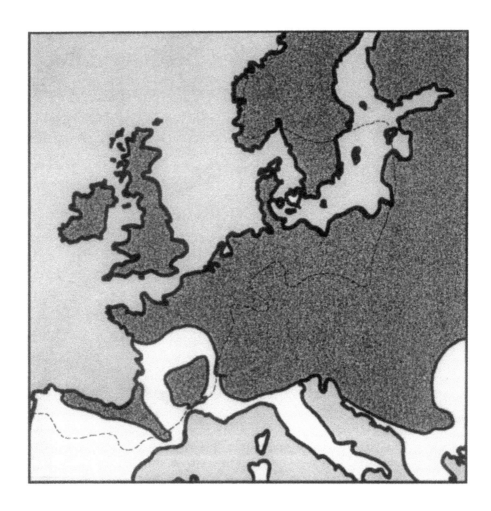

Lacerta vivipara.

Family Lacertidae

Podarcis muralis

Common Wall Lizard (GB), Lézard des murailles, Mauereidechse (DE), Muurhagedis (NE), Lagartija Roquera (ES), Murödla (SE).

Introduction

This species was first described in 1768. The name *Lacerta muralis* was adopted in 1802 and used until the 1970s. The species was then put in the Genus *Podarcis*. For years the Wall Lizard has been known from its habit of using walls and in 1901 Gadow referred to it as follows: "The Walled Lizard deserves its name, since in the Mediterranean countries there is scarcely a wall on which these active lizards do not bask or run up and down, often head downwards, in search of insects".

Taxonomy

There are many species of small lizards in Europe (collectively called Small Lacertids). For convenience they have been broadly divided into two groups: the viviparous, rock and meadow lizards; the wall lizards. The so-called Common Wall Lizard is concentrated on here. The generic name is a compound world derived from Greek and intended to mean sure or certain foot, that is sure-footed. Many sub-species have been described and probably the best source of information about the complexities of deciding what species have been accepted is Bohme's Reptilien un Amphibien Europas. They are often quite extraordinarily difficult to identify. However in some localities, such as the north of Spain, where some live in sympatry, the external morphology of each species, *Podarcis muralis, P. hispanica, P. bocagei*, can easily be recognised.

Protection

This lizard is listed in Appendix II of the Berne Convention. Some countries such as the Czech Republic, have designated protected areas for the conservation of the species. It is also included on lists of protected species in many countries including Belgium, the Czech Republic, France, Hungary, the Netherlands, Spain and Switzerland.

Description

This small lacertid grows to about 7 cm (s-v) in length and the tail may be twice as long as the body. When active, it has a characteristically flattened and delicate appearance. The scales are slightly keeled and there is usually a small collar. Its colour is delightfully varied both within and between populations. Essentially it is brownish or grey with faint amounts of pale green and typically it has black and white bars on the sides of the tail.

Distribution and habitat

The Wall Lizard is distributed throughout Europe as far north as France and south to Italy and central Spain. It is found on some of the islands off the Spanish Cantabric coast and is present on the Channel Islands. Wall Lizards have been introduced to England where there are now thriving colonies such as on the Isle of Wight.

As its name suggests, it does frequent walls where there is much sunlight but of course that is a man-made habitat. It is basically very adept at climbing and the species should be regarded as inhabiting places where there are warm, sheltered banks, rock faces, scree slopes and trunks of trees. It generally remains within a short distance from a refuge, and in one study that distance was found to be 0.8 m. At high altitudes this species has been found to inhabit river valleys. Because of its alert and opportunistic nature it has become well adapted to living in and around buildings, in fact almost entirely on some buildings. The most northern natural population is in Maastricht in the Netherlands where it is found on the remains of ancient fortifications.

Seasonal movement and behaviour

The Wall Lizard is diurnally active for approximately 255 days of the year and its daily cycle is influenced by weather conditions. In northern localities – near Bonn – unusually long activity seasons from March to November have been recorded

Vagility and population ecology

The population size and density of this species has been found to depend on the complexity of the habitat, and in some localities there may be more than 90 animals in a colony. It has been found that about 40% of individuals do not remain in the colony but tend to move between colonies. Individuals have been found to move distances of between 10 and 90 m and there is one record of a male moving over 70 m in as little as 90 minutes.

Feeding ecology

The diet of this small lizard is extremely varied. Insects, especially flies but also true bugs, bees, wasps, earwigs, beetles and grasshoppers, and amphipods, spiders, worms and molluscs are all taken. It also seems that it will eat small fruits. Spiders are particularly common in young lizard diets. The ratio of insect material to other items of food can be as high as 85%. Smaller proportions of insects occur in the diet in some localities but only because of particularly high numbers of non-insect taxa in those places.

Thermal ecology

This lizard is a typical diurnal heliotherm. Morning emergence is temperature dependent and occurs between 8.00 and 9.00 hours. In southern localities, the daily activity pattern is bimodal with diminished activity during the hottest part of the day. In northern parts this species inhabits large stony and rocky areas where there are exceptionally warm microclimates. For normal activity, it maintains body temperatures within a 33 to 36° C range, and such precision can be achieved by shuttling between sun and shade. The mean body temperature of lizards kept in vivaria is between 33 and 34° C. In the wild, the body temperature of both males and non-gravid females are similar, but those for gravid females are lower because they stay closer to refuges and do not exploit thermal conditions to a maximum.

Reproduction, growth and development

In some locations there can be up to three times as many females as males, but the sex ratio is usually about equal. There is a considerable difference in the timing of the reproductive period throughout the species range. In France, for example, mating occurs from March to mid-April but in Germany and the Netherlands it may occur as late as mid-June. Aggressive behaviour and fighting between males commonly occurs during the mating period. Eggs are laid in soil about four weeks after mating but, depending on the region, laying of eggs may occur from the end of April to the middle of August. Clutch sizes vary between two and ten eggs and it is known that some females may lay two or even three clutches within one year. The incubation period is temperature dependent and may vary between six weeks and as much as five months, but is typically between six and ten weeks. A temperature of 28° C has been found to be optimum for incubation in terms of embryo development rate, hatching success and post-hatching performance.

General

Further work on the colouration and factors related to the different colour forms – green versus brown – may help to clarify the taxonomy of this species.

Major references

Avery, R.A. (1978) Activity patterns, thermoregulation and good consumption in two sympatric lizard species (*Podarcis muralis* and *P. sicula*) from central Italy. Journal of Animal Ecology, 47, 143-158.

Barbault, R. & Mou, Y.P. (1988) Population dynamics of the common wall lizard, *Podarcis muralis*, in southwestern France. Herpetologica, 44, 38-47.

Boag, D.A. (1973) Spatial relationships among members of a population of wall lizards. Oecologia, 12, 1-13.

Brana, F. (1991) Summer activity patterns and thermoregulation in the wall lizard, *Podarcis muralis*. Herpetological Journal, 1, 544-549.

Brana, F. (1993) Shifts in body temperature and escape behaviour of female *Podarcis muralis* during pregnancy. Oikos, 66, 216-222.

Dexel, R. (1986) Zur Okologie der Maurereidechse *Podarcis muralis* (Laurenti, 1768) (Sauria: Lacertidae) an ihrer nordlicher Arealgrenze. Salamandra, 22, 63-78.

Galan, P. (1986) Morfologia y distribucion del Genero Podarcis, Wagler, 1830 (Sauria, Lacertidae) en el Noroeste de la Peninsula Iberica. Revista Espanola de herpetologia, 1, 87-142.

Saint Girons, H. & Duguy, R. (1970) Le cycle sexuel de Lacerta muralis L. en plaine et en montage. Bulletin Museum d'histoire naturelle, Paris, 42, 609-625.

Strijbosch, H., Bonnemayer, J.J.A.M. & Dietvorst, P.J.M. (1980) The northernmost population of *Podarcis muralis* (Lacertila, Lacertidae). Amphibia-Reptilia, 1, 161-172.

Podarcis muralis. **Male (a), female (b).**

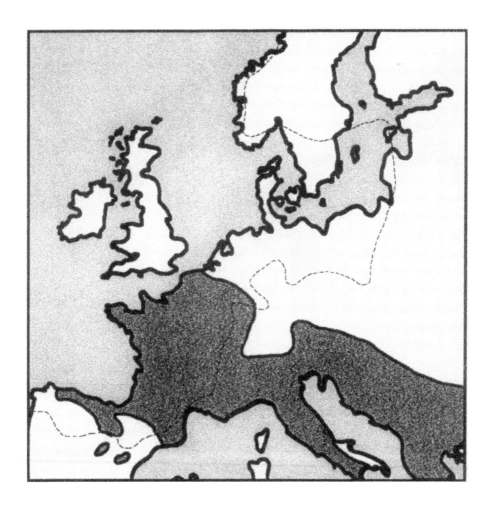

Podarcis muralis.

Family Colubridae (Typical snakes)

Coronella austriaca

Smooth Snake (GB), Coronelle lisse (FR), Glattnatter oder Schlingnatter (DE), Gladde slang (NE), Culerbra Lisa Europea (ES), Hasselsnok (SE).

Introduction

The Smooth Snake was not reported in Britain until late in the 19th century. In the summer of 1853 a Mr Bond and the Revd Cambridge collected a specimen at Dorset. Although at the time it was thought to be a new species for Britain, that specimen was preserved and forgotten. Six years later the scientific journal, *Zoologist*, carried a brief 11-line article by a Mr Gray, which reported the Smooth Snake as an addition to Britain's fauna. Later, in 1886, Revd Cambridge wrote a detailed account accompanied by a fine line drawing for the Proceedings of the *Dorset Natural History* and *Antiquarian Field Club.*

Taxonomy

The delightful name *Coronella* refers to the crown-like mark or stripe on the animal's head. The specific name *austriaca* is used because the first scientific description came from Austria in 1768. The name 'smooth' refers to the smooth, soft scales that lack a central keel, unlike those of the Adder and Grass Snake. There is a closely related and very similar species *Coronella girondica*, the Southern Smooth Snake, and one disputed sub-species, *C.a. fitzingeri*, from the south of Italy and Sicily.

Protection

This snake is listed in Appendix II of the Berne Convention and is fully protected in most countries where it occurs.

Description

This species has a small head, poorly defined neck and small eyes with round pupils. Its colour is variable but usually brown, reddish-brown or greyish with small dark blotches and spots extending down the length of its back and tail. A dark stripe along the side of the neck extends across the eye to the nostril. The male and female are very similar except that the male's tail is longer. In England adult Smooth Snakes weigh between 40 and 50 g and their average body length is between 50 and 60 cm, with some reaching 75 cm.

Distribution and habitat

This snake has been recorded from southern England and Scandinavia, the former USSR and south to Spain and Italy. In southern England recent records indicate that it is more widespread in the counties, Hampshire and Dorset than was previously thought. In southern England, the Smooth Snake is found on heathlands and in habitats adjacent to or recently derived from heaths. In north-west Europe the three main components of its habitat seem to be soil and litter in which it can burrow, a dense ground vegetation in which it can thermoregulate, and an upper layer of scrub or woodland in which it can forage safely for food. In the south it may be found in rocky places.

Seasonal movements and behaviour

Smooth snakes seem to spend much of their time hidden beneath the ground or in dense, low-lying vegetation. It is therefore difficult to follow their movements throughout the year. The general pattern seems to be emergence from the wintering site during March, mating during May and June, birth of young in September or October, and retreat to a wintering site by October. All this activity is temperature dependent and there may be wide variations from year to year. They over-winter beneath the ground where they are protected from the frost and cold weather. They emerge as early as mid-February but most seem to do so towards the latter half of April and after the other two species of snakes have commenced their summer activity. The nature of the autumn weather affects when the last Smooth Snakes retreat to their winter refuges, but it is usually by early October.

Vagility and population ecology

The Smooth Snake's behaviour is erratic and it tends to remain within small areas with distances traveled in a day often no more than 15m. On some occasions it may travel over 100 m in a day. It does not have a territory because it is not known to defend an area but it does tend to remain in the same place or home range. The estimated average area (uncorrected for sample size) of the home range in habitats of mixed forest and heathland has been found to be 9,690 square metres, and that for open heathland 985 square metres. When those values are corrected for sample size the home range size is increased several fold.

Its distribution seems to have become very fragmented over the past few decades and it has probably become extinct in Denmark. Its secretive nature makes it difficult to determine population densities and impossible to ascertain or even estimate total numbers. Those estimates that have been made are misleading and worthless because so many assumptions have been made.

There is no reliable way of locating all individuals in a population and, whereas estimates of the number of Smooth Snakes in Britain, for example, are pure conjecture, there have been useful calculations of some population densities, based on extensive field studies. These vary from 0.9 snakes per hectare (forest-heathland habitat) to 1.9 snakes per hectare (open heathland). Higher population densities have been reported for Smooth Snakes but it is certain that these are all under-estimates, because it was impossible to find all the animals.

An example of how difficult it is to estimate population sizes of Smooth Snakes arose out of a study on a 25 ha national Nature Reserve. Smooth Snakes were captured and individuals marked over a period of several years. Captures of new individuals had become uncommon, suggesting that a large proportion of the population had been seen. In 1976 a fire destroyed much of the reserve but some live and dead Smooth Snakes were found amongst the vast area of ashes. Surprisingly those snakes found after the fire were on average larger than any previously established average size, indicating they were old animals that had not been caught previously. One interpretation is that only a small proportion of Smooth Snakes in any population will be found and those are more likely to be the larger individuals.

Feeding ecology

The distribution of Smooth Snakes is similar to Sand Lizards, except in the south of Europe, resulting in the fallacy that these snakes are major predators of those lizards. Smooth Snakes do occasionally catch small Sand Lizards but various species of lizard and mammal are taken. In one survey, 41 Smooth Snakes took 63 prey items, 15 of which were Common Lizards and the rest were rodents and other small mammals of which nestling rodents and shrews seemed to be common.

Thermal ecology

Smooth Snakes spend less time basking than do Adders and Grass Snakes. The reason seems to be that Smooth Snakes have lower body temperature preferences. When studied in the laboratory, it was found that the average body temperature of a Smooth Snake was 28° C, three degrees lower than an Adder's. The body temperature of Smooth Snakes in the laboratory, however, have been found to be higher than those in the field and this observation suggests that for some reason they are not able to exploit heat sources as effectively as Adders.

Reproduction, growth and development

Female snakes usually breed every second year and the gestation period of three to five months is very much affected by temperature. Periods of fine weather may result in early births around August but unfavourable conditions may hinder development resulting in births being delayed till October. There have been some cases where females have remained gravid over the winter and given birth early in the following spring. At birth Smooth Snakes are less than 20cm in total body length and weigh between 2.2 and 3.8 g. This weight is doubled in the first year. They are not sexually mature until at least three years and population studies suggest a longevity of at least ten years. There are some records of marked Smooth Snakes surviving for as long as 19 years. Few young snakes are ever seen, suggesting a high mortality in the first few years of life – or they are just extremely elusive. The mortality and restricted movement of this snake seem to make it very vulnerable to habitat change and fragmentation, and to isolation.

General

Smooth Snakes are secretive in their habits and even the most experienced field worker may find only one animal after hours of searching. The best time of year for field studies is either during April and May, or in the autumn, particularly during warm weather following a period of rainfall, when adults may often be seen basking on embankments or forest ride verges in the late afternoon or early evening sunlight. This species has been well researched but there is much room for studies on the effectiveness of habitat restoration and management.

Major references

Andren, C. & Nilson, G. (1976) Hasselsnoken (*Coronella austriaca*) – en utrotningsotad ormart! Fauna och flora, 2, 61-76.

Bont, R.G.de, Gelder, J.J.van & Olders, J.H.J. (1976) Thermal ecology of the Smooth snake, *Coronella austriaca* Laurenti, during spring. Oecologia, 69, 72-78.

Gent, A.-H. (1988) Movement and dispersion of the Smooth Snake *Coronella austriaca* Laurenti in relation to habitat. 529 pp. Ph.D. thesis, University of Southampton.

Goddard, P. (1984) Morphology, growth, food habits and population characteristics of the Smooth snake *Coronella austriaca* in Southern

Britain. Journal of Zoology, London, 204, 241-157.

Luiselli, L., Capula, M. & Shine, R. (1996). Reproductive output, cost of reproduction, and ecology of the smooth snake, *Coronella austriaca,* in the Eastern Italian Alps. Oecologia 106, 100-110.

Spellerberg, I.J. & Phelps, T.E. (1977) Biology, general ecology and behaviour of the snake *Coronella austriaca.* Biological Journal of the Linnean Society, 9, 133-164.

Volkl, W. & Meier, B. (1988) Verbreitung und Habitatwahl der Schlingnatter *Coronella austriaca* Laurenti, 1768 in Nordostbayern. Salamandra, 24, 7-15.

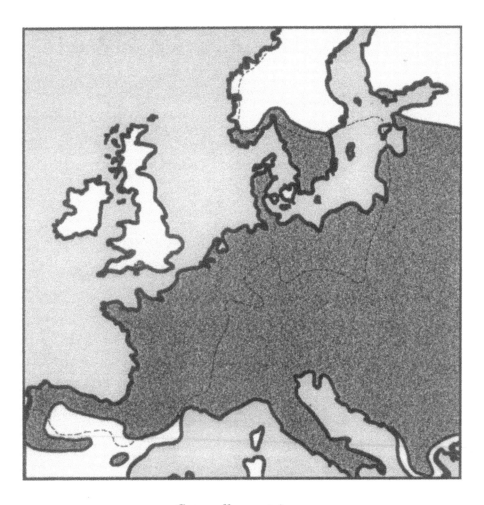

Coronella austriaca.

Coronella austriaca. **Adult.**

Family Colubridae

Natrix natrix

Grass or Ring Snake (GB), Couleuvre à collier (FR), Ringelnatter (DE), Ringslang (NE), Culebra de Collar (ES), Snok, Vanlig snok, or Vattensnok (SE).

Introduction

This is probably the most well-known snake in temperate Europe after *Vipera berus*. Around the turn of the 20th century, reports of plagues of *Natrix natrix* were not uncommon. In 1900, for example, there was a report of 1,200 Grass Snakes which were found in a Welsh village and later killed. Its size has also been the subject of many reports. There is much variation in the appearance of Grass Snakes with several sub-species being recognized. This account is largely about the Western Grass Snake *Natrix natrix helvetica*.

Taxonomy

The name *Natrix natrix* refers to its aquatic habits and means water serpent. The geographic variation in the Grass Snake has been well researched and it is clear that there are two main forms, the Eastern and Western Grass Snakes. There are also some island forms. In central Europe, the Eastern form meets the Western along a 'hybrid' or transition zone. Nine sub-species have been described.

Protection

There is a lack of information about the status of this timid species but it is protected in most countries. In Britain it is illegal to sell or harm Grass Snakes. Anecdotal reports suggest that, in general, population levels have declined in the UK and that elsewhere its distribution has diminished. Paradoxically there have been reports of increases in abundance, possibly due to the construction of new ponds that are well stocked with amphibians. Although difficult to obtain, information about *Natrix natrix* populations is urgently needed. Monitoring schemes, as established in the Netherlands, to provide an index of change in populations, would be useful.

Description

This species has a well-defined head, its eye pupil is round, and its scales are very strongly keeled. The females are larger than males and may reach lengths of up to 120 cm or even more. The colour of *Natrix natrix* is varied and that variation can be made to appear more prominent depending on where the snake has been and over what kind of ground it has been travelling. Excursions over wet mud or clay, for example, can result in rather strangely coloured snakes. In general they are basically olive-green in colour but this may vary from olive-grey to olive-brown – black individuals have occasionally been reported, particularly from Scandinavia. There are black spots and vertical bars along the length of the animal. A notable, although variable, feature is the collar, the orange, yellow or white marks at the base of the head. The colour patterns of the collar seem to vary from region to region. The underside is characteristically checkered with black and white or black and grey – sometimes entirely black.

Distribution and habitat

The Western Grass Snake has a widespread distribution and is found across north Africa, Iberia, the Italian peninsula, France, Britain and parts of Germany and Switzerland. It is also found on a number of islands including Jersey. The sub-species *N. n. natrix* found in Scandinavia occurs as far as 66° North. In Britain, *N. n. helvetica* occurs as far north as the Scottish border.

This snake is typically found in damp places, such as wet meadows, around standing water and along the banks of streams. It is sometimes found, however, a long way from any standing water and seems at home in wooded localities and on heathland. On occasions these snakes will bask openly in grassy areas but their preferred habitat seems to be scrub with an abundance of brambles. They are very agile swimmers and may quickly avoid capture by slipping into water and darting out of sight.

Seasonal movements and behaviour

In the north of its range, this snake has been found to be strictly diurnal, even on warm summer nights. Towards the south of its range there is evidence that is may move about at night. One of the first snakes to be seen in spring, its behaviour when disturbed is characteristically noisy as it hurries away through the vegetation. Male Grass Snakes are very active, particularly during the mating season when looking for females; gravid females similarly so when searching for incubation sites. Both sexes are most commonly seen basking during the spring after which their behaviour seems quite unpredictable. Towards the end of October they seek winter refuges which may possibly be used by many snakes. Grass Snakes can be active throughout mild winters.

Vagility and population ecology

Field research on *Natrix natrix* in the 1970s showed that they were not at all easy to study because of their unpredictable and sometimes wide-ranging movements across the countryside. In some studies, many snakes were individually marked but the rate of recapture or resighting was always very low. More recent research using implanted miniature radio transmitters has confirmed that it does move about the countryside in an unpredictable manner. Distances in straight lines between sitings of 600 to 700m a day and up to one km over two to three days have been recorded. The actual distance traveled is of course much more than this because the animals do not travel in straight lines. They do not seem to have a home range although some do appear to move within fairly well-defined areas. Hedgerows seem to be well used as linear habitats and possibly as corridors between other kinds of habitat.

Feeding ecology

N. natrix forages widely and feeds on aquatic vertebrates such as fish, frogs, toads and newts. It also eats young birds and small mammals. The most popular prey appears to be toads, which is surprising because of their chemical defenses.

Thermal ecology

Grass snakes, like other snakes, are ectothermic and must spend time basking to raise their body temperature to a level at which normal body functions are most efficient. On warm sunny days Grass Snakes are able to raise their body temperature very rapidly to the mean voluntary temperature of 30° C. They do move about actively, however, with body temperatures of between 16 to 18° C.

Reproduction, growth and development

N. natrix lays eggs. It seems that large females may lay between 30 and 40 eggs during June and July, possibly as many as 50; young females as few as 10 eggs. Particularly interesting is the fact that this species will actively seek egg-laying sites which will help incubate the eggs: compost heaps, dung heaps, piles of sawdust or chippings from saw mills, piles of leaves and other decaying mounds of vegetation are commonly used. It is also thought that some snakes could be attracted to the warmth of baking ovens in the houses of old villages, and hence the possibility that the Welsh village mentioned above could have attracted so many female snakes.

The eggs which have parchment-like shells, are about 1.8 cm in breadth and 2.8 cm in length when first laid but increase a little in size during incubation because of absorption of moisture. The rate of incubation is temperature dependent and may take as long as 10 weeks in cool conditions or about 6 weeks in warm ones.

Hatching, that is cutting through the parchment-like skin of the egg, is facilitated by a fairly prominent egg-tooth – common among lizards and snakes. This is not a tooth but a tiny sharp projection lying in front of the upper jaw which is shed soon after hatching. The newly-hatched young are 15 to 18 cm in length and reach about 28 to 30 cm after the first year.

General

This species has not been well researched and the following may prompt some ideas for future investigations. Grass Snakes are particularly sluggish when their body temperature is low and are thus vulnerable to predators such as raptors, corvids and mammals such as hedgehogs and mustelids. Annual mortality rates have been found to be lower in immature, rather than mature Grass Snakes and this difference seems to be associated with changes in reproductive behaviour. The high rate of movement and activity of mature males therefore seems to make them more susceptible to predation. There is conflicting information about clutch size and factors determining clutch size. Mortality of eggs and hatchlings seems rather high and there are some reports of deformed (often double-headed) young Grass Snakes.

Major references

Brown, P.R. (1991) Ecology and vagility of the Grass Snake, *Natrix natrix helvetica* Lacepede. 197 pp. Ph.D. thesis, University of Southampton.

Gaywood, M.J. (1990) Comparative Thermal Ecology of the British Snakes. 182 pp. Ph.D. thesis, University of Southampton.

Luiselli, L., Capula, M. & Shine, R. (1996). Food habits, growth rates and reproductive biology of grass snakes *Natrix natrix* (Colubridae) in the Italian Alps. Journal of Zoology (London), 241, 371-380.

Madsen, T. (1983) Growth rates, maturation and sexual size dimorphism in a population of grass snakes, *Natrix natrix*, in southern Sweden. Oikos, 40, 277-282.

Madsen, T. (1984) Movements, Home Range Size and Habitat Use of Radio-tracked Grass Snakes (*Natrix natrix*) in southern Sweden. Copeia, 1984, 707-713.

Madsen, T. (1987) Cost of reproduction and female life-history tactics in a population of grass snakes, *Natrix natrix*, in southern Sweden. Oikos, 49, 129-132.

Thorpe, R. S. (1975) Quantitative handling of characters useful in snake systematics with particular reference to intraspecific variation in the ringed snake *Natrix natrix* L.. Biological Journal of the Linnean Society, 7, 24-43.

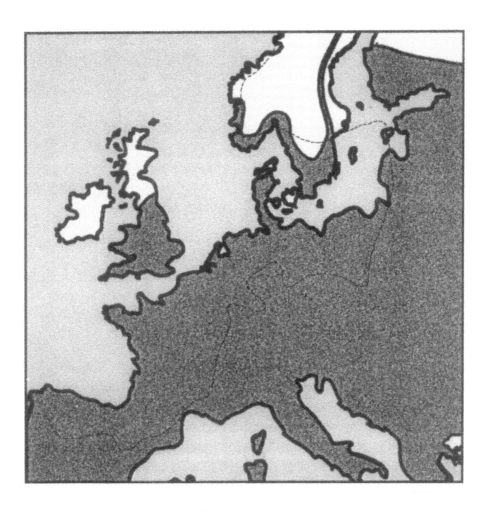

Natrix natrix.

Natrix natrix. Adult.

Family Viperidae

Vipera berus

Adder or Viper (GB), Vipère péliade (FR), Kreuzotter (DE), Adder (NE), Vibora Europea (ES), Huggorm (SE).

Introduction

The name Viper is derived from two Latin words, *vivus* 'alive' and *pario* 'to bring forth'. In parts of Europe the name Adder is derived from the Saxon nedre meaning 'lower', alluding to the snake's creeping position (the same source as for the Natterjack Toad). The literature, both ancient and modern, has many references to vipers but not all are *Vipera berus*, and indeed there are seven species of viper found in parts of Europe. The ability of Adders to swallow and protect their young in the face of danger, then regurgitate them, was a topic much debated in the early literature to the extent that one author devoted an entire chapter to the subject. Adders cannot swallow then regurgitate their young and this is one of the many delightful fallacies associated with the species.

Taxonomy

The specific name *berus* was at one time used for the Generic term for a snake. A few sub-species have been described: the nominate sub-species *Vipera berus berus* is found in most of Europe, the sub-species *V. b. bosniensis* is found in the mountainous regions of the Balkan peninsula. There is some doubt about the status of *V. b. seoanei* from the north-west of the Iberian peninsula but it is usually given the full species status, and indeed two sub-species have been described. The new-born and young Adders tend to be a lovely brick-red colour, leading to the suggestion that this was a different species; indeed at one time the name *Vipera rubra* was suggested but later rejected.

Protection

For *V. berus*, as with other venomous species, there can be an interesting conflict between the need to protect a species which is declining and the perception that it is dangerous. Protected by law in some countries, for example Belgium, Denmark, Germany, Hungary, the Netherlands, Norway and Switzerland, there has been much criticism of that protection. In Britain when it was decided to protect Adders from being killed as well as from being sold – under the Wildlife and Countryside Act 1981, amendment 1991 – there were some unjustified critical and cynical letters in the national press. Recent use of its venom to help monitor levels of radiation in Russia could possibly improve the Adder's image.

Description

Adders are thick-bodied, stocky snakes which reach about 65 cm in total length, though commonly achieve about 50 to 55 cm. Female Adders are larger than males. They have vertical pupils in the eyes, and dorsal body scales are strongly keeled. The head and neck markings vary and are specific to individuals – a good way to identify and study the behaviour of them. Their colour provides a particularly interesting, rare example of sexual dimorphism found among snakes. Adult females are usually brownish or reddish-brown with dark brown markings, whereas the males are commonly a contrasting black and pale grey with a rather obvious zig-zag line on their dorsal surface. The zig-zag pattern and their colour makes them particularly difficult to see amongst dead bracken fronds.

Totally black (melanism) Adders occur occasionally, and on an island off Sweden about 50% of males were found to be melanistic. There has been an interesting debate about the advantages of blackness. This colour may make them more visible, for example, and thus vulnerable to predation. If a black Adder absorbs heat more effectively it might be thought to grow more quickly and reproduce earlier. Possibly the black colour is relatively insignificant in terms of heat because an increase in body temperature is a function of the total reflectance of the animal's skin, not just the colour (see Fig 6). No difference has been found in weight and length between melanistic and normally coloured Adders.

Distribution and habitat

The Adder has the largest geographical range of any terrestrial snake species is Europe. It occurs patchily through much of Europe, except Spain, as far north as the Arctic Circle. It is also found in some localities in the former USSR as far as the Pacific coast. In parts of France populations are found adjacent to Vipera aspis but the two species are mutually exclusive.

Adders can be locally common and occupy a wide range of habitats from forest and woodlands through heathlands and moors to hedgerows and embankments associated with agricultural land.

Seasonal movement and behaviour

Adder behaviour throughout the year appears to be synchronized with emergence and the first moult occurring at the same time in local populations. It is usually the first snake to appear in the spring with males emerging first though not moving far from their winter refuges. Communal wintering appears to occur but it is not unusual to find several males basking together in early March once daytime temperatures are regularly above 8° C. Females emerge some weeks later but move beyond the males towards the areas where they will spend the summer.

Both visual and chemical cues are used in male-male encounters. If more than one male locates a receptive female, the fascinating combat dance between males may occur, with each snake rising up and attempting to push the other to the ground. This combat behaviour seems to be induced by pheromones produced by the sexually active females and is stimulated by visual cues from rival males. Territorial defence in the Adder is maintained only for a few weeks during the reproductive period and the territory defended is the immediate area around a reproductive female.

Vagility and population ecology

Adders have a distinct seasonal pattern of movement but individuals may remain in the same locality and use precisely the same basking site for several weeks. To some extent the movements and basking behaviour is more predictable than in other snakes. The population density can be quite varied, for example in the Swiss Alps, there may be about 3 adults per hectare and the home range about 5.2ha for males and 0.7 for reproductive females.

Feeding ecology

Rodents predominate in the Adder's food and frogs, fledgling birds and lizards are also taken. In Poland the most common prey was found to be the Bank Vole *Clethrionomys glareolus*, and in England it was the Short-tailed Vole *Microtus agrestis*. In a unique population study of Adders off the coast of Sweden, the prey density of rodents was found to affect the population of Adders: when prey abundance fell, there was a significant decrease in the mean body mass of both females and males. At the same time, only the body length in males increased. The survival of non-reproductive females was higher after one year of low prey densities.

Thermal ecology

Adders have the highest voluntary temperatures of all snakes in the European temperate climatic region, and their behavioural thermoregulation is well adapted to maximize heat uptake. They bask for long periods and their mean voluntary body temperature is 33° C. The critical minimum temperature has a seasonal variation from 3.0 to 0.5° C, an obvious advantage in cold climates.

Reproduction, growth and development

Adders, like other snakes, have a Jacobson's sense organ in the roof of the mouth which is highly sensitive to chemical particles (pheromones). The flicking tongue is used to collect and then transfer chemical signals to Jacobson's organ and by this method the male Adders are able to follow chemical trails left by the females or by prey. Body contact (tactile behaviour) also seems to be an important component of behaviour during the reproductive season and during courtship; body contact is maintained almost continuously.

In Adders, as in some other snake species, copulation triggers ovulation, which then occurs in late April. In May, after copulation, males may be found 'guarding' the female. Although other male snakes may succeed in mating with the same female, there has been a suggestion that they will not sire any young because of a copulatory plug which forms after the first mating and prevents additional sperm from entering the uterus. There is some doubt whether multiple mating in Adders increases female reproductive success through genetic benefits.

Female Adders breed every second year and sometimes every three years. They usually give birth during August or early September. If the summer is exceptionally cold, birth may not take place till the following spring.

General

In Britain the last detailed ecological study of Adders was undertaken during 1959-61. Since that time there have been many anecdotal reports of increase in Adder populations during hot summers. Sadly we have little information on which to determine the population status of this much maligned species.

Major references

Andrén, C. (1986) Courtship, mating and agonistic behaviour in a free-living population of adders, *Vipera berus* (L.). Amphibia-Reptilia, 7, 353-383.

Andrén, C. & Nilson, G. (1983) Reproductive tactics in an island population of Adders, *Vipera berus* (L.), with a fluctuating food source. Amphibia-Reptilia, 4, 63-79.

Nilson, G. (1981) Ovarian cycle and reproductive dynamics in the female adder, *Vipera berus* (Reptilia, Viperidae). Amphibia-Reptilia, 2, 63-82.

Parker, G.A. (1992) Snakes and female sexuality. Nature, 355, 395-396.

Pomianowska-Pilipiuk, I. (1974) Energy balance and food requirements of adult vipers *Vipera berus* (L.). Ekologia Polska, 22, 195-211.

Presst, I. (1971) An ecological study of the viper *Vipera berus* in southern Britain. J. Zool, Lond., 164, 373-418.

Saint Girons, H. (1975) Coexistence de *Vipera aspis* et de *Vipera berus* en Loire-Atlantique: un probleme de competition interspecifique. Terre et la Vie, Rev. d'Ecologie Appliquee, 29, 590-613.

Viitanen, P. (1967) Hibernation and seasonal movements of the viper, *Vipera berus* berus (L.) in southern Finland. Annales Zoologici Fennici, 4, 472-546.

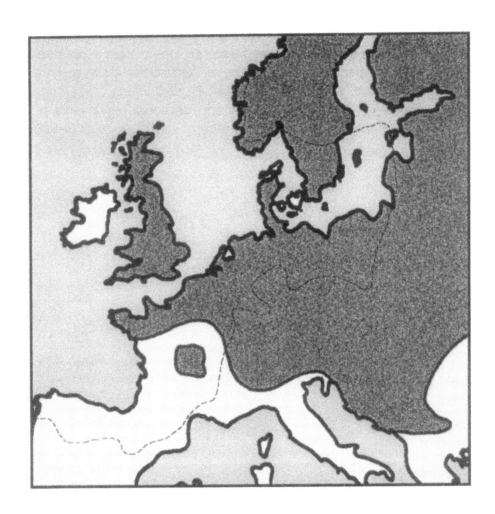

Vipera berus.

Vipera berus. **Male (a), female (b).**

Chelonia

Tortoises, terrapins and turtles

The Order Chelonia includes the tortoises (terrestrial), the terrapins (aquatic or semi-aquatic) and the turtles (marine). Some species of tortoises and terrapins have been imported in large numbers to Britain in the past. The frequency of turtles around the coastline seems to have increased in recent years, and some frequently find their way to this part of Europe.

Tortoises and terrapins

The three European tortoises occur in a wide range of habitats but are generally restricted to areas with hot dry summers. Hermann's Tortoise *Testudo hermanni* has no spurs on its thighs whereas the Spur-thighed Tortoise *Testuda graeca* is aptly named. The third species, Marginated Tortoise *Testudo marginata*, has a very distinctive long carapace with very flared plates at the rear, thus the specific name. The Generic name *Testudo* is Latin for tortoise. Hermann's Tortoise is named after Professor J. Hermann (1738-1800), a professor of natural history at Strasbourg. The Common European Tortoise inhabits large areas of southern Europe despite its specific name graeca, of Greece.

These tortoises are protected under the Berne Convention but their habitats are being destroyed at an alarming rate. In 1990 the site of the largest known concentration of Hermann's Tortoise on mainland Greece was destroyed. About 3,000 animals may have been destroyed as the habitat was flattened for building developments. As well as large-scale habitat loss, tortoise populations have been exploited. For example, hundreds of thousands of tortoises and terrapins have been collected and imported to parts of northern Europe to be sold as pets. The huge numbers which were imported to Britain legally (with a licence) reached a peak in 1967 with the staggering figure of 419,876. Not surprisingly this exploitation has devastated many of the natural populations. Few of the imported tortoises in Britain survived, and the mortality has been estimated at 90%. What of the survivors? One possible logical step to take would be to establish an organization for the safe repatriation of all remaining tortoises in Britain and thus make a sensible effort to try and restore the exploited populations. This would make sense as long as suitable habitat remains and the safety of the repatriated tortoises can be ensured.

Terrapins, particularly the European Pond Terrapin *Emys orbicularis*, have been imported to northern Europe in far lower numbers than tortoises. There are occasional sightings and reports of terrapins in Britain, some of which have been deliberately released and others have escaped from captivity. Terrapins have a more flattened appearance than tortoises. They are web-footed and have a long tapering tail. Very agile swimmers, European Pond Terrapins inhabit well-vegetated, still or slow-flowing water where they feed on a wide variety of aquatic animals.

Marine Turtles

There are a few species of marine turtles which occur as rare vagrants on the coasts around Britain, France, Belgium and Germany. Although they are all excellent swimmers, individual turtles have strayed from their usual environments in warmer seas. World-wide there are seven species of marine turtles and there are confirmed records for five of these off temperate coasts of north-west Europe. Turtles are threatened by pollution, hunting and particularly by the loss of nesting beaches. Furthermore, climate change may have disastrous effects. The sex of turtles is determined by temperature (see introduction) and in increase in temperature could result in only one sex being produced. Sea level rises could also threaten nesting beaches. All species are threatened and trade in turtle products is banned in many countries under the Convention in International Trade in Endangered Species (CITES). All turtles are fully protected in British waters under the Wildlife and Countryside Act.

Dermochelys coriacea
Leathery Turtle

The Generic name comes from derma skin and khelus tortoise, while the specific name comes from corium leather. The Leathery Turtle or Leatherback is a regular visitor to our shores with several records each year, and has been recorded as far north as Shetland and the Norwegian Sea. These impressively streamlined turtles may reach 2 m or more in length and can weigh over 750 kg. They are black or dark brown with white spots. One large specimen found dead on a beach in the west of England was a sad reminder of the perils of marine pollution. This animal which was 1.6 m in length and weighed about 315 kg, had died after swallowing a plastic bag containing fishing tackle. Another one weighing 202 kg, drowned off the coast of Northumberland after becoming entangled in lobster-catching gear. Leathery Turtles are carnivorous, feeding mainly on jellyfish and other small marine organisms. Their dark leathery skin covers a bony carapace which has five to seven prominent ridges.

Caretta caretta

The name *caretta* is derived from carey meaning tortoise-shell. This species breeds in the Mediterranean, where it is at great risk from the ever-expending developments on the beaches, and also in the Atlantic, Pacific and Indian Oceans. It is smaller than Dermochelys with a carapace of about 100 cm in length covered in reddish-brown or olive-coloured horny plates. Loggerheads feed on jellyfish, crabs, shellfish and other small marine creatures.

Rare visitors

Three other species of marine turtle occasionally wander from their usual more tropical environments and are seen in the coastal waters of Britain and north-west Europe. The smallest is the grey/olive green Kemp's Ridley Turtle *Lepidochelys kempii*, which has a carapace no greater than 65 cm in length, covered in black, grey or olive-coloured plates with small keels on the dorsal mid-line. The Hawksbill Turtle *Eretmochelys imbricata*, with a prominent hooked beak, grows to about 90cm in length and has a reddish-brown, yellow and black, beautifully marked carapace. The old tortoise-shell ornaments were made from the horny scutes on the Hawksbill's carapace. The Green Turtle *Chelonia mydas*, which reaches 1.4 m in length, is a very rare visitor from tropical waters around Latin America, Africa and Asia. Similar to the Loggerhead, the Green Turtle is dark olive brown in colour and can be distinguished by the four, instead of five, plates on each side of the carapace.

Major references

Committee on Sea Turtle Conservation, National Research Council (1990) Decline of the Sea Turtles. National Academy Press.

Lambert, M.R.K. (1984) Threats to Mediterranean (West Palearctic) tortoises and their effects on wild populations: an overview. Amphibia-Reptilia, 5, 5-15.

Meek, R. (1985) Aspects of the ecology of *Testudo hermanni* in southern Yugoslavia. British Journal of Herpetology, 6, 437-445.

Willemsen, R.E. & Hailey, A. (1989) Status and conservation of tortoises in Greece. Herpetological Journal, 1, 315-330.

Testudo hermanni. **Adult**

Emys orbicularis. **Adult.**

Dermochelys coriacea. **Adult.**

Caretta caretta. **Adult.**

Conservation of Amphibians and Reptiles

Threats to species and habitats

As with many other forms of wildlife, pollution, pesticides, exploitation and habitat loss and damage have had their effects on amphibian and reptile populations (Table 1). Habitat damage and loss is the main reason – but other contributing factors include road kills (Hels and Buchwald, 2001; Reh and Seitz (1990) and the indirect effect of climate change. The exploitation of crocodiles and turtles has been well documented and so also has the distasteful trade in frogs for food (frogs' legs).

For hundreds of years, by way of example, frog populations in Europe have been exploited for the French and Belgian markets, and today there is such a demand for frogs' legs that populations elsewhere are used to satisfy that demand. In 1990, 6,202 tonnes of frogs' legs were imported into the European Community (Patel, 1993). Once the main source, India has now been replaced by Indonesia as the largest supplier of frogs' legs to Europe. Consequently some species may soon become extinct and the important role of frogs as a natural predator of insect pests has diminished.

In most cases the impact of collecting amphibians and reptiles for private study and private collections has also had an impact on wild populations, particularly the rare species. Regrettably the trade in these animals continues throughout the world and rare and endangered species are taken from the wild and sold to collectors. The practice of taking amphibians and reptiles and keeping them in captivity for enjoyment should have disappeared with the old type of zoo. In Britain, the Great Crested Newt, the Natterjack Toad and the Sand Lizard are just three species which have dramatically declined over the last few decades, and loss of habitat and collecting are two contributing factors.

Value

Does it matter that so many amphibians and reptiles are declining in numbers and that some have become extinct in recent years? Do they have any 'value'? Some species are or have been of some economic importance and a few have been the source of medicinal

Table 1. Reasons for the decline in amphibian and reptile populations.

Climate change
Exploitation for food, materials and trophies
Pollution of water
Drainage of wetlands
Deforestation (loss of native forests)
Excessive water extraction and drying up of rivers and ponds
Afforestation with alien species (intensive forestry)
Development of land for agriculture
Pesticides
Extraction of minerals
Construction of transport routes
Road kills
Tourism
Urbanization
Collecting for the pet trade
Release of non-native predators
Persecution

materials. The value of wildlife need not be based on monetary values; there are alternative ways of perceiving the worth of animals and plants (Spellerberg, 1992). They can be seen, for example, as being a valuable part of our natural or even cultural heritage. They could be an integral part of an ecosystem. Many species of amphibians and reptiles are sensitive to environmental change and can be regarded as valuable biological indicators, that is indicators of deteriorating environmental condition (Spellerberg, 1991). The declining populations of some species can provide an early warning of pollution and change in the state of the environment. Perhaps most importantly the case for improved conservation of amphibians and reptiles can be made on the grounds that they have received a disproportionately small amount of attention in the past as regards conservation when compared to other vertebrate groups.

Conservation

What is being done to conserve amphibians and reptiles? There are educational programmes, some species are protected by laws, nature reserves have been established for some species and habitat restoration and creation has been undertaken in some countries. Conservation strategies have in particular been developed for human dominated landscapes (Bray and Gent, 1997).

Herpetological conservation groups such as the Conservation Committee of the Societas Europaea Herpetologica have done much to promote the need for conservation education programmes, as has the British Herpetological Society through the efforts of its own Conservation Committee (Corbett & Tamarind, 1979: Corbett, 1988). More recently the Herpetolical Conservation Trust has done much to co-ordinate conservation activitites in the United Kingdom.

Many European countries now have legislation protecting amphibians and reptiles (see Corbett, 1989, for a comprehensive account). In mainland Britain The Wildlife and Countryside Act 1981 and subsequent amendments is the most important piece of legislation (it is also the means by which the Berne Convention is implemented). Schedule 5 of that Act lists animals which are protected: some fully, some to the extent that they may not be sold, others to the extent that they may not be killed or injured. Every five years the Schedules are revised. In the 1991 revision of Schedule 5, the further protection of the Adder (with respect to killing and injuring) caused much wry comment in the national press.

The Berne Convention on the Conservation of European Wildlife and Natural Habitats, which came into force in 1982, is a particularly important piece of wildlife legislation for Britain and Europe. Developed through the Council for Europe, the aims are to conserve wild flora and fauna and their natural habitats, to promote cooperation between countries in their conservation efforts and to give particular attention to migratory, endangered and vulnerable species. This Convention has lists (appendices) of plant and animal species for which various levels of protection have been agreed. Appendix 2 lists those animals in need of special protection, such as the Smooth Snake *Coronella austriaca*.

The European 'Habitats Direction', the Convention on the Conservation of European Wildlife and Natural Habitats, is the EU's attempt at implementing the Berne Convention through member states, and it draws attention to the conservation needs of certain species in various countries (Table 2).

Table 2. Recommendations for amphibian and reptile species protection in Europe and Scandanavian countries. (European "Habitats Directive").

Species	Countries
Bufo calamita	Austria, France, The Netherlands, Sweden, UK
Hyla arborea	Belgium, Denmark, Germany, Luxembourg, The Netherlands, Sweden, Switzerland
Triturus cristatus	Austria, Belgium, Finland, France, Germany, Luxembourg, The Netherlands, Norway, Sweden, Switzerland, UK
Lacerta agilis	Belgium, Denmark, France, Germany, Greece, Luxembourg, The Netherlands, Spain, Sweden, UK

Note: Countries are recommended to pay particular attention to the species mentioned above. That does not mean that these species require no attention in the countries not mentioned, nor does it mean that those species not mentioned are secure. (Source: Recommendation Nos. 26 & 27 of the Standing Committee of the Council of Europe Convention on the Conservation of European Wildlife and Natural Habitats).

The restoration and management of habitats has been a particularly important component of the conservation of amphibians and reptiles. Information from research has provided a basis for the design of good conservation programmes, and now is benefiting species such as the Natterjack Toad *Bufo calamita* in England and the Moor Frog *Rana arvalis* in both The Netherlands and Germany. In the Netherlands, woodland is considered to be important in conjunction with pools and pool age can affect species numbers (Laan and Verboom, 1990).

A prerequisite for conservation is good information about the distribution and populations of the relevant species. In Britain, English Nature, Scottish Natural Heritage, Countryside Council for Wales and the Joint Committee for Nature Conservation have sponsored and published data on the distribution and status of amphibians and reptiles, including: The Ecology and Conservation of Amphibian and Reptile Species Endangered in Britain (1983); Cooke & Scorgie (1983) The Status of the Commoner Amphibians and Reptiles in Britain; Brown & Oldham (1990) The Status of the Widespread Amphibians and Reptiles in Britain. English Nature has published a Priority Natural Area for mammals, reptiles and amphibians which includes conservation pritorities for selected species.

The success of habitat restoration and habitat management can be assessed most effectively by well-designed monitoring programmes. Indeed habitat restoration and management without any follow-up studies would seem difficult to justify. Good ecological monitoring requires standardised and regular collection of data over a period of time which may extend to several years (Spellerberg, 1991). The methods of storage of the information can be equally important and so also can the need to ensure continuity of the programme with the possibility of changes of staff.

Surely methodologies are integrated elements of monitoring. Reading (1996, 1997) undertook a three-year study to evaluate methods used. His report provides an essential reference for the best survey methods. Foster and Gent (1996) also contribute to the important topic of survey methods.

Fortunately throughout Europe there are surveys and monitoring programmes which provide the much-needed information on the status and distribution of amphibians and reptiles (see for example the Atlas of European reptiles and amphibians). In Britain there are many county recording schemes for these animals and also comprehensive national surveys. For example, Swan and Oldham, based at De Montfort University, have directed a national Amphibian Survey and a national Common Reptile Survey (Swan & Oldham, 1989 & 1993). Many of these surveys contributed to country distribution maps and material recording schemes such as the United Kingdom and the Biological Records centre and the National Biodiversity Network.

Threats from exotic species

Some amphibians and reptiles found in Europe are not indigenous species. A few of these exotic species have and continue to cause problems. In Germany for example the American Bullfrog (and other alien species) have been declared unwanted exotics. There has been some concern that such alien species may displace indigenous species.

Some European Studies in Conservation

Excellent European examples of conservation programmes for species of amphibians and reptiles include those based on attempts to reintroduce species, those based on the important task of surveys, on detailed ecological research and on good examples of collaborative ventures with industry. The varied nature of such work can be seen in the following examples.

Sweden

The National Council for the Protection of Nature and the Swedish WWF have supported herpetological conservation research programmes which have resulted in a number of habitat restoration projects. The amphibians of Sweden have been particularly in need of habitat restoration and reintroduction programmes because of the effects of 'acid rain' which, over the last decade, have resulted in increased levels of acidity in ponds, lakes and streams.

Claes Andrén and Göran Nilson have undertaken much of the conservation work on reptiles and amphibians in Sweden. For many years they have been directing research projects on several species of reptiles, especially *Lacerta agilis*, *Coronella austriaca*, *Vipera berus* (see references under species accounts). Their work on the amphibians has involved some typical temperate region species such as *Bufo calamita* and *Rana arvalis*. They have also undertaken ecological research on amphibians that have become extinct from the temperate region of Sweden, for example a long-term programme on the toad *Bombina bombina* (Andren & Nilson, 1979), the first amphibian species to become extinct. Closely related to *Bombina variegata*, *B. bombina* is today found only on the very edge of the temperate climatic region of Europe. The aim of this particular programme is to re-establish viable populations of this toad within the next ten years.

During the last 100 years the Swedish populations of *B. bombina* decreased from being 'fairly common', with 21 documented breeding sites, to extinction in 1960. It was in 1982 that the Swedish authorities approved a proposal by Andren and Nilson to reintroduce the species. The first phase of the project involved research on reasons for its decline. Localities where it was previously found had undergone dramatic changes both in water quality and the surrounding land use. In many cases the ponds were surrounded by intense agriculture where there would only be localities available for over-wintering.

The basic information about the biology of *B. bombina* and its habitat requirements were assembled from the literature and from discussions with colleagues who had undertaken research on the species. The criteria used for reintroduction sites were as follows:
1. Suitable local microclimates with high daily mean temperatures;
2. At least two, but preferably five, suitable shallow water bodies all within migration distance (500m);
3. Water free from fish, with plenty of submerged vegetation and a varying water table;
4. Surrounding land with balanced grazing pressure and no use of agricultural chemicals;
5. The protection of the area should have been established for a long period of time.

With no Swedish animals available for use, Andren & Nilson maximized genetic variation from as many populations as possible in nearby Denmark. Over three years, a total of 900 eggs were collected from many female toads in three regions. During the first three years most of the eggs were cultured and the toads were released in three areas. The reintroduction involved 70 potential breeding ponds and in as many as 26 ponds about 70 80 males have been heard calling. Successful reproduction has now been confirmed in up to ten ponds.

Switzerland

Amphibian populations have experienced a substantial and continuing decline over the past few decades in Switzerland, and today most species are considered to be threatened (Grossenbacher, 1994). As a first step towards establishing conservation programmes, it is important to have information on the distribution and status of the species. Since 1988, new legislation has allowed the Swiss Federal Government to designate habitats of national importance and to oblige the cantons (States) to protect these sites. To help prevent the continuing decline of amphibian populations in Switzerland, a group of herpetologists was mandated in 1989 to prepare an inventory of amphibian breeding sites of national importance. This is the Koordinationsstelle fur Amphibien und Reptilienschutz in der Schweiz (KARCH) (Jan Ryser, pers, comm.).

All breeding sites in the country were assessed (nearly 8,000) using known data on species richness, population sizes, and the rarity of a species as a whole as well as on a regional basis. About ten per cent of the highest-ranking sites were to be included in the inventory, and these were studied in 1990 and 1991. Field workers defined three zones requiring varying degrees of protection level. These three zones were as follows:

A.   a core area which includes the actual breeding ponds and surrounding wetlands;
B.   the surrounding zone which includes neighbouring agricultural land or woods;
C.   gravel and clay pits and similar habitats which, although artificial, do play an important part in conservation strategies.

The inventory contains 891 breeding sites, averaging 7.00 ha in size (median 1.8 ha). Before the inventory can be acted upon, the cantons are consulted. The Federal Government considers the inventory an important contribution towards implementing the Berne Convention.

In addition to this important survey and census work, there has been a major voluntary effort to reduce the extent or mortality of amphibians on roads. These efforts have involved both drift fences and pitfall traps – which direct the animals to certain points alongside the road where they can be collected – and road tunnels (popularly called toad tunnels). The KARCH has records of 49 localities with toad tunnels and there are about 148 sites where temporary drift fences are erected (Jan Ryser, pers. comm.).

There are also species specific conservation programmes in Switzerland. For example, in three cantons, efforts have been made to conserve *Hyla arborea*. In one canton, Aargau, the private nature conservation organisation Schweizerischer Bund fur Naturschutz, has purchased an area of land where attempts to increase the population density of the species are being made. Former agricultural land in these areas is being converted to wetlands by removing a layer of soil and by creating shallow pools. Similar work is being undertaken in areas on the border with Austria.

The Netherlands

The amphibians and reptiles in Holland have long been the subject of detailed surveys, and consequently much is known about the distribution and status of all species. Databases of this information have been created and they provide an important aspect of the monitoring of these animals. Some notable projects are on the snake *Natrix natrix* and on providing more and better habitats for amphibians.

Conservation has for many years been associated with detailed behavioural and ecological research, much of the latter being attributed to Henk Strijbosch and his colleagues and students since the late 1970s (see the reference under species accounts and general reading). Information about habitat selection by nine species of amphibians (Strijbosch, 1980), for example, resulted from detailed research in an area of inland dunes covered mostly with woods and fens. Since 1972, Strijbosch and his students have carried out very detailed ecological research on *Lacerta agilis* with particular emphasis on its reproductive biology (Strijbosch, 1988). The fact that this species lays eggs and leaves them to the elements is perhaps the part of its life cycle, which endangers so many Sand Lizard populations. It has been particularly important, therefore, to obtain detailed information which can provide a basis for ensuring a greater availability of suitable, safe nest sites.

Germany

The number of literature references from Germany which have been cited at the end of the species accounts does not do justice to the amount of ecological research and the extent of conservation efforts directed at amphibians and reptiles in that country. The 'Hanbuch der Reptilien und Amphibien Europas' (Bohme, 1984) is but one testament to the concerted herpetological research interests in Germany.

In this small space it is also not possible to do justice to the concern about declines in amphibians and reptiles not only in Germany but also throughout the world. For example, several research groups in Germany have drawn attention to and undertaken research on the status of amphibian spawning sites and environmental factors affecting the survival of eggs and larvae.

The distribution, ecology and behaviour of many species of amphibians and reptiles has been researched in Germany for many decades and more recently certain species have been targeted for special attention. It has been those special, concerted efforts, which illustrate the extent of collaborative work, which has so often been undertaken.

For example in the 1980s, two symposium reports were published as a result of two very successful meetings which brought together the expertise from several countries including Germany; one meeting was directed at the Moor Frog (*Rana arvalis*) and the other was directed at the Sand Lizard (*Lacerta agilis*). The reports from those meetings were edited by Dieter Glandt and Richard Podloucky (Moor Frog) and by Dieter Glandt and Wolfgang Bischoff (Sand Lizard). Full references to the reports may be found in the respective species accounts.

If one species only were to be noted for the extensive research underlying conservation and management in Germany, it would be *Lacerta agilis*. In addition to the Symposium noted about (by Glandt & Bischoff) there has been extensive ecological research undertaken by Andreas Nollert on the subspecies *L. a. argus*. His 1989 publication 'Beitrage zur Kenntnis der Biologie der Zaundidechse, Lacerta *agilis argus* (LAUR.), dargestellt am Beispiel einer Population aus dem Bezirk Neubrandenburg (Reptila, Squamata: Lacertidae)' includes remarks on habitat management and conservation based on a six year study. Some detail from this research and the full reference to the publication is given in the species account.

United Kingdom

The following are two case studies of conservation in practice, which took place in the south of England under quite different circumstances. The first was an attempt to conserve a population of Sand Lizards Lacerta agilis following a severe fire on a National Nature Reserve – a reactive form of conservation. The second arose from mitigating work undertaken by BP Exploration when faced with problems of developing an onshore oilfield, which lies beneath herpetologically important areas of heathland.

Conservation of *Lacerta agilis* following destruction of habitat by fire

In 1976 a fire destroyed much of a heathland National Nature Reserve on which there were populations of Sand Lizards and other reptiles (Spellerberg & House, 1982). A few lizards survived but were at risk from predators in the blackened landscape without any vegetation. A total of 30 lizards was subsequently collected from the burnt heathland and kept temporarily in a small enclosure.

The decision was later taken to try and resettle the animals on a small part of the Nature Reserve untouched by fire. It was considered important to monitor such a resettlement programme. For this reason a large enclosure (18 x 30m) was constructed and the enclosed area was modified on the basis of results from previous detailed ecological research. Wintering and basking sites wee especially constructed, the former being sheltered cavities below ground and the latter being logs partly covered by overhanging vegetation.

For a few years after the release of the lizards in the enclosure, insects and spiders were collected from nearby heathland and deposited in the enclosure. In 1981, the enclosure contained a small breeding population of lizards and the vegetation on nearby previously burnt areas has regenerated. In 1982 the enclosure was dismantled and the lizards dispersed into the heathland. Since that time the animals have continued to increase in numbers and recolonize many parts of the Nature Reserve.

Restoration of herpetological sites by BP Exploration

The following is a brief account of some of the work undertaken by BP Exploration to minimize herpetological disturbance in the development of the Wytch Farm oilfield. All work described was carried out under licence from the then Nature Conservancy Council, now English Nature (Reading, 1991).

Wytch Farm Wellsite

The Wytch Farm oilfield is made up of a number of wellsites interconnected by pipelines to a central gathering station. One, a heathland site, was unusual in that it had already been developed as a temporary wellsite and had, over a number of years, been relatively unmanaged with the result that heathland vegetation, including conifers up to 2 m high had recolonised the gravel area surrounding the wellhead. Prior to upgrading the area to a permanent concrete, sealed wellsite in 1989 BP commissioned the Institute of Terrestrial Ecology (ITE) Furzebrook Research Station to survey the site for reptiles and amphibians and then, as a result of the survey, conduct a capture and relocation programme for the animals that were present.

Three methods were used to catch reptiles and amphibians. Reptiles were either searched for by eye and captured by hand or after being attracted to artificial refuges placed within the wellsite area. As the large freshwater storage pond within the boundary area was drained, hand-netting was used to catch amphibians, Palmate Newts *Triturus helveticus*, and a number of large eels that were present.

Captured reptiles were released on suitable heathland at least 600 m away from the wellsite and its surrounding land management area and separated from it by a conifer plantation, thus reducing the likelihood of animals finding their way back to the wellsite area. Captured newts and other aquatic life were released into a nearby watercourse with similar habitat.

With the exception of the Adder *Viperus berus*, all the remaining endemic British reptile species were found on the wellsite area. Of these the Sand Lizard *Lacerta agilis* was most abundant, followed by the Common Lizard Lacerta vivipera and Slow Worm *Anguis fragilis*. A total of 64 reptile and 46 amphibian captures were made at this site in 1989.

In 1990, following wellsite development, very few reptiles were seen or captured on site. The decline in the number of reptile sightings (20) indicated that the reptile removal programme carried out during 1989 had been a success. During 1992, the number of reptile sightings within the land management area retained for landscaping and screening of the wellsite had increased to 72 with the Slow Worm and Sand Lizard being the most abundant species.

The increase in reptile sightings within this area was undoubtedly due, in part, to the relative lack of disturbance at the site. This allowed reptiles from the adjoining heath to recolonize the heathland retained within the land management area that surrounded the wellsite.

Additionally an important aspect of the programme was the mitigation made at improving the habitat for reptiles. While wellsite development was under way a large south-facing sandbank at the northern end of the site was returfed with blocks of heather retrieved from elsewhere on site. Although initially slow to recover the transplants were successful despite a long, hot, dry spring and summer.

Purbeck – Southampton pipeline (PSP): Wareham Forest

Oil extracted from the reservoirs is piped from the wellsites to a central gathering station at Wytch Farm where it is treated before being transported to the Hamble oil terminal on Southampton Water. Due to the large volumes of oil being extracted an oil pipeline (PSP) was laid between the gathering station and the Hamble oil terminal. This pipeline passed through Wareham Forest and at one site an area known to be inhabited by all six British reptiles was crossed. BP Exploration commissioned ITE Furzebrook Research Station, Wareham, to undertake a capture and removal programme for reptiles along the 500 m section of affected mature heathland and to monitor its subsequent recolonization following pipe-laying operations.

After an initial short capture and release programme it became apparent that it would not be feasible to catch all the reptiles present in the area due to the height of the heather: the abundance of 'bold holds' and the time constraints of the project. A scheme therefore was devised to make the area unattractive to reptiles so that they would leave of their own accord, and this ultimately proved very successful. This scheme was achieved in stages:

1.   the numerous small conifers (1-3 m) were removed using chainsaws and bow saws.
2.   Initially a group of men with 'strimmers' formed a line and walked along the PSP route reducing the vegetation height to approximately 5-10 cm. This method, however, proved to be less effective than using a large rotary blade attached to the rear of a tractor (swipe), as adapted later. As the tractor progressed forward along the working width of the PSP route it was closely followed by observers looking for reptiles. Although the rotary blade reduced the vegetation height to a few centimetres no reptiles were killed. Indeed one male Sand Lizard was picked up unharmed after the tractor had passed over it.
3.   Most of the cut vegetation was removed so that the area offered little or no cover for reptiles.
4.   The area was then left for a number of weeks before the heavy pipe-laying machinery was brought in.
5.   An additional factor in making the area unattractive to reptiles was the degree of noise and disturbance that was created during the vegetation clearance operations.

Once the pipeline had been laid the area was leveled, seeded and/or turfed, and the heathland vegetation allowed to recolonize it. Monitoring the recolonization of the PSP route by heathland plants and animals (reptiles) was undertaken over an initial 3-year period. This was achieved using large numbers of artificial refuges, which both reptiles and amphibians are attracted to, on both the PSP route and the adjoining heathland. During the first year (1989) only one adult female and one hatchling Sand Lizard were seen on the PSP route, although a number of Sand Lizard sightings were made along the boundary of the PSP with the adjacent heath. All five of the remaining British reptile species were seen on the heath adjoining the route. One amphibian, *Bufo bufo*, was regularly found under refuges on the PSP route but not on the adjoining heath.

During 1991 both Sand Lizards and Smooth Snakes *Coronella austriaca* were seen periodically on the PSP route. Grass snakes *Natrix natrix*, which had occurred on the PSP route in 1990 were not encountered there in 1991 but were seen on the adjoining heath. Only the Slow Worm, Common Lizard and Adder were seen exclusively on the adjoining heath and not on the PSP route. Although relatively few reptile sightings occurred on the PSP route between 1989 and 1991 evidence was found to suggest that it did not form a barrier preventing reptiles from crossing it.

Under present British legislation, developers have an obligation to minimize the potentially deleterious effects that any form of development may have on some endemic species of reptile and amphibian. To this end detailed Environmental Statements are required to determine the scale of any potential problem. If protected species are found to occur within proposed development areas the expert advice must be sought over how best to deal with the problem. Advice and expertise are available from English Nature, the ITE, the Herpetological Conservation Trust and the British Herpetological Society. Indeed, English Nature by law has to be approached as it has responsibility for issuing licences allowing the specially protected species to be handled.

Finally, forward planning is essential when dealing with sites that contain breeding populations of protected reptiles or amphibians. Due to the life cycle of these animals it is often necessary to begin approaching the problems at least 6 to 12 months before development of an area commences so that enough time is allowed for discovering hibernating, breeding and foraging sites which may be widely separated. In addition, if trap and release programmes are required sufficient time must be allowed for them to be completed.

Glossary

Abiotic (factors)   These are non-biological or environmental factors such as temperature, light intensity, humidity, etc.

Acclimation   The process of physiological adjustment which occurs under laboratory conditions.

Acclimatization   The process of physiological adjustment to changing seasonal temperature levels, pH, light intensity, etc., which occurs under natural conditions.

Allopatric   If two or more species have an allopatric distribution, their distribution is geographically dissimilar.

Arboreal   Living in trees.

Aquatic   Living in water.

Biome   A biological region classified on the basis of climate and the major vegetation type, e.g., tropical rainforest biome, taiga – coniferous forest biome.

Circadian   A period of activity which is about 24 hours long.

Cline   A graded series of different forms of the same species.

Crepuscular   Active during sunrise or evening (twilight).

Critical minimum   In amphibians and reptiles, the lower body temperature at which locomotion is no longer possible.

Diel   During the daily (day and night) cycle.

Diurnal   Active during the hours of daylight.

Endogenous   In reference to rhythmic behaviour, it is that behaviour which has a rhythm based on an organism's biochemical cycles – cf. Exogenous.

Eutrophication   The nutrient enrichment of bodies of water caused by organic enrichment. Although a natural process, rapid eutrophication can use up oxygen in the water and this may result in the mortality of the aquatic organisms.

Euryoecious   A species that is able to tolerate a wide range of conditions or is not specialised, e.g., an eurythermal species of frog is one which may tolerate a wide range of water temperature conditions (cf. stenoecious).

Exogenous   Refers to rhythmic behaviour which is based on environmental

	cycles such as changes in the seasons, tides and cycles of day and night.
Fecundity	The capacity of an individual to multiply (breed): for example, the number of eggs produced gives a level of fecundity.
Genetic (variability, variation)	Refers to the external variation in appearance of individuals of the same species.
Gestation	Period of development of an embryo.
Habitat	the locality or area used by a population of organisms and the place where they live.
Hibernation	The specialised physiological state which occurs in some mammals during the winter resting period.
Impervious	refers to water not being able to permeate or pass across on through the skin.
Metamorphosis	the change in form of an animal as it goes from the larval stage to the adult form.
Microhabitat	Habitat which is very small.
Neoteny	A state found amongst some amphibian species in which larval characteristics are retained (may be caused by environmental conditions), cf. Paedomorphosis.
Niche	The space occupied and the resources used by a species. Conceptually the niche has many dimensions and each resource used by the species – type of food, period of activity, water depth, etc. – can be considered as one dimension. If a species is not a specialist or can be considered to use a wide range of food and conditions then such a species would be considered to have a wide niche. The niche width can be calculated mathematically.
Nocturnal	Active during hours of darkness.
Optimal foraging	The feeding behaviour which uses least effort for gaining the most food in least amount of time.
Paedomorphosis	A state found amongst some amphibians in which the adult mature form retains larval characteristics including external gills.
pH	An indication of the acidity or alkalinity (base) of a material expressed on a scale with neutral at pH 7. Lower values indicate acidity and higher values alkalinity.
Pheromone	A chemical, or group of chemicals, released by an organism which in very small concentrations is used either as an attractant or an alarm signal.
Reproductive isolation	Circumstances in which different species living in the same area will not interbreed because of some biological or behavioural mechanism.

Resource partitioning	The nature by which species living in the same area share the resources of food, space, time (of activity), etc.
Riparian	Occurring along the side of a watercourse such as a stream, river or canal.
Sexual dimorphism	The morphological features of the sexes which are dissimilar, e.g. Adders (*Vipera berus*) have a sexual dimorphism.
Species	A group of organisms of the same kind that can reproduce sexually among themselves and do not usually breed with other, related organisms or other species. It is the basic taxonomic unit, which has two names, the Generic and the specific. In some instances, species can be divided into sub-species, geographic races and varieties.
Spermatophore	A small gelatinous container (packet) containing sperm.
Stenoecious	A species which has narrow tolerances or is very specialised.
Sympatric	A form of distribution where two or more species share the same area – cf. Allopatric.
Thermoregulation	The regulation of body temperature.
Vagility	Refers to the rate of movement and extent of movement of a species. A highly vagile species is one, which moves around frequently and over long distances.
Xeric	Dry, arid. A species with adaptations to arid conditions.

General Reading and References Cited in the Introduction

Andrén, C. and Nilson, G. (1979) The Fire-bellied Toad (*Bombina bombina*) and its return to Sweden. Fauna och Flora, 74, 253-270.

Arnold, E.N. (1987) Resources partitioning among lacertid lizards in southern Europe. J. Zool. Lond. B, 1, 739-782.

Arnold, E.N. and Burton, J.A. (1978 & 1992) A Field guide to the Reptiles and Amphibians of Britain and Europe. 272 pp Collins, London.

Barbadillo, L.J. (1987) Guia de Incafo de los anfibios y reptiles de la Peninsula Iberica, Islas Baleares y Canarias. 694 pp. INCAFO, Madrid.

Beebee, T.R.C. (1983) The Natterjack Toad. 120pp. Oxford University Press, Oxford.

Beebee, T.R.C. (1985) Discriminant analysis of amphibian habitat determinants in South-East England. Amphibia-Reptilia, 6, 35-43.

Beebee, T.R.C. and Griffiths, R. (2000). Amphibian and Reptiles. A natural history of the British Herpetofauna, 270 pp Harper Collins, London.

Bohme, W. (1984) Reptilien und Amphibien Europas. Pp. 416 Aula, Wiesbaden.

Boulenger, G.A. (1847) The Tailless Batrachians of Europe. 2 vols. 210 & 376 pp. The Ray Society, London.

Bray, R. and Gent, T. (eds) 1997. Opportunities for amphibians and reptiles in the designed landacape. Proceedings of a seminar at Kew Gardens, Richmond, Surrey 24 January, 1996. British Nature Science, 30, 82 pp.

Coates, M. and Ruta, M. 2000. Nice snakes, shame about the legs. TREE,15,503-504.

Cloudsley-Thompson, J.L. 1999. The Diversity of Amphibians and Reptiles. 254 pp. Springer, Berlin

Cooke, M.C. (1865) Our Reptiles. 199 pp. Harwicke, London. Revised edition (1893) 200 pp. W.H. Allen, London.

Corbett, K.F., ed. (1989) The Conservation of European Reptiles and Amphibians. 274 pp. Christopher Helm, London.

Ellis, E.A. (1979) British Amphibians and Reptiles. 32 pp. Jarrold, Norwich.

Foster, J. and Gent. (eds). 1996. Reptile survey methods: proceedings of a seminar 7 November 1995 at the Zoological Society of London's meeting rooms, Regent's Park, London. English Nature Science 27, 223 pp.

Gadow, H. (1901) Amphibia and reptiles. The Cambridge Natural History. 280 pp. Macmillan, London.

Gaywood, M.L. (1990) Comparative Thermal Ecology of the British Snakes. 182 pp. Ph.D. thesis, University of Southampton.

Goddard, P. (1981) Ecology of the Smooth Snake, *Coronella autriaca* Laurenti in Britain. 180 pp. Ph.D. thesis, University of Southampton.

Griffiths, R. 1996. Newts and Salamanders of Europe. 188pp. Academic Press, London

Grossenbacher, K. (1994) Rote Liste der gefahrdeten Amphibien der Schweiz. In Duelli, P. ed., rote Listen der gefahrdeten Tierarten der Schweiz. Pp. 33-34. Buwal – Reihe Rote Listen.

Gruber, U (1989) Die Schlangen Europas. 248 pp. Frankh-Kosmos, Stuttgart.

Günther, R. (1990) Die Wasserfrosche Europas. 288 pp. Ziemsen, Wittenberg.

Halliday, T. & Arano, B. (1991) Resolving the Phylogeny of the European Newts. Trends in Ecology and Evolution, 6, 113-117.

Hels, T. and Buchwald, E. 2001. The effect of road kills on amphibian population. Biological Conservation, 99, 331-340.

Helmich, W. (1962) Reptiles and Amphibians of Europe. 160 pp. Blandford, London.

Kalezic, M.L. & Hedgecock, D. (1980) Genetic variation and differentiation of three common European newts (*Triturus*) in Yugoslavia. British Journal of Herpetology, 6, 49-57.

Laan, R. and Verboom, B. 1990. Effects of pool size and isolation on amphibian communities. Biological Conservation, 54. 251-262.

Langton, T. 1989. Snakes and Lizard. 125pp. Whittet Books, London.

Latreille, P.A. (1800) Histoire naturelle des salamandres de France. 61 pp. Crapelet, Paris.

Lizana, M., Perez-Mellado, V. & Ciudad, M.J. (1990) Analysis of the structure of an amphibian community in the central system of Spain. Herpetological Journal, 1, 435-446.

Mann, W., Dorn, P. & Brandl, R. (1991) Local distribution of amphibians: the importance of habitat fragmentation. Global Ecology and Biogeography Letters, 1, 36-41.

Mertens, R. & Wemuth, H. (1960) Die Amphibien und Reptilien Europas. 264 pp. Kramer, Frankfurt.

Nöllert, A. (1984) Die Knoblauchkrote. 103 pp. Ziemsen, Wittenberg.

Nöllert, A. & C. (1992) Die Amphibien Europas. 382 pp. Frankh-Kosmos, Stuttgart.

Oldham, R.S. & Nicholson, M. (1986) Status and ecology of the Warty Newt (Triturus cristatus). Unpublished report by Leicester Polytechnic (now De Montfort University) for the Nature Conservancy Council (now English Nature). NCC CSD Report No. 703.

Patel, T. (1993) French may eat Indonesia out of frogs. New Scientist, 138 (1868), 10 April, 7.

Reading, C.J. (1991). Recolonisation of disturbed heathland by reptiles and amphibians. Institute of Terrestrial Ecology Annual Report 1990-1991. 48-50

Reading, C.J. (1996). Evaluation of reptile survey methodologies: final report. English Nature Research Reports 200, 48pp

Reading, C.J. (1997). A proposed standard method for surveying reptiles on dry lowland heath. Journal of Applied Ecology, 34, 1057-1069.

Reh, W.L., and Seitz, A. 1990. The influence of land use on the genetic structures of populations of the Common Frog *Rana temoraria,* Biological Conservation, 54, 239-249.

Rocek, Z. (1986) Studies in Herpetology. Proceedings of the European Herpetological Meeting (3rd Ordinary General Meeting of the Societas Herpetologica) 1985. 754 pp. Charles University, Prague.

Roesel von Rosenhof, A.J. (1758) Historia naturalis ranarum nostratium. Die naturliche Historie der Frosche hiesigen Landes, folios of different lengths. Nurenberg. (in Latin & German).

Salvador, A. (1985) Guia de campo de los anfibios y reptiles de la Peninsula Iberica, Islas Baleares y Canarias. 255 pp. Garcia, Leon.

Schreiber, E. (1875) Herpetologia Europaea. 639 pp. Vieweg, Braunschweig.

Simms, C. (1970) Lives of British Lizards. 128 pp. Goose, Norwich.

Slater, F. (1992) The Common Toad. 20pp. Shire, England

Smith, M. (1951) The British Amphibians and Reptiles. 322 pp. Collins, London.

Spellerberg, I.F. (1976) Adaptations of reptiles to cold. In: Bellairs, A. d'A. & Cox, C.B., eds., Morphology and Biology of Reptiles. Linnean Society Symposium Series. Pp. 261-285. Academic Press, London.

Spellerberg, I.F. (1982) Biology of reptiles. An ecological approach. 158 pp. Blackie, Glasgow.

Spellerberg, I.F. (1991) Monitoring Ecological Change. 334 pp. Cambridge University Press, Cambridge.

Spellerberg, I.F. (1992) Evaluation and Assessment for Conservation. 288 pp. Chapman & Hall, London.

Spellerberg, I.F. & House, S.M. (1982) Relocation of the lizard *Lacerta agilis*: an exercise in conservation. British Journal of Herpetology, 21, 921-933.

Steward, J.W. (1969) The Tailed Amphibians of Europe. 180 pp. David & Charles, Newton Abbot.

Steward, J.W. (1971) the Snakes of Europe. 238 pp. David & Charles, Newton Abbot.

Strijbosch, H. (1979) Habitat selection of amphibians during their aquatic phase. Oikos, 33, 363-372.

Strijbosch, H. (1980) Habitat selection by amphibians during their terrestrial phase. British Journal of Herpetology, 6, 93-98.

Swan, M.J.S. & Oldham, R.S. (1989) Amphibian Communities. Unpublished report by Leicester Polytechnic (now De Montfort University) to Nature Conservancy Council (now English Nature).

Swan, M.J.S. & Oldham, RS. (1983) Herptile Sites. Vo.1, National Amphibian Survey Final Report. Research Report No. 38. Vol.2, National Common Reptile Survey Final Report. Research Report No. 39. English Nature, Peterborough.

Wisniewski, P.J. (1989) Newts of the British Isles. 20 pp. Shire, England.

Recordings

In addition to written reports there are some very useful recordings of vocalizations of amphibians. For example:

Roche, J. –C. & Guyetant, R. Frogs and Toads. The calls of 20 species found in France, Belgium and Switzerland. Sittelle, Chateaubois, 38350 La Mure, France.

Index

For Product Safety Concerns and Information please contact our EU
representative GPSR@taylorandfrancis.com
Taylor & Francis Verlag GmbH, Kaufingerstraße 24, 80331 München, Germany

www.ingramcontent.com/pod-product-compliance
Ingram Content Group UK Ltd.
Pitfield, Milton Keynes, MK11 3LW, UK
UKHW050925180425
457613UK00003B/29